Genstat 5
A Second Course

PETE DIGBY

Statistics Department
AFRC Institute of Arable Crop Research
Rothamsted Experimental Station, Harpenden

NICK GALWEY

Department of Applied Biology, University of Cambridge

and

PETER LANE

Statistics Department
AFRC Institute of Arable Crop Research
Rothamsted Experimental Station, Harpenden

CLARENDON PRESS · OXFORD
1989

Oxford University Press, Walton Street, Oxford OX2 6DP
Oxford New York Toronto
Delhi Bombay Calcutta Madras Karachi
Petaling Jaya Singapore Hong Kong Tokyo
Nairobi Dar es Salaam Cape Town
Melbourne Auckland

and associated companies in
Berlin Ibadan

Oxford is a trade mark of Oxford University Press

Published in the United States
by Oxford University Press, New York

British Library Cataloguing in Publication Data
Digby, Pete
Genstat. 5.
1. Statistical analysis. Applications of computer systems
I. Title II. Galwey, Nick III. Lane, Peter, 1952–
519.5'028'553
ISBN 0–19–852201–0
ISBN 0–19–852218–5 (pbk.)

Library of Congress Cataloguing in Publication Data
Digby, P. G. N.
Genstat 5.
Bibliography: Includes index.
1. Genstat (Computer system) 2. Mathematical statistics—Data processing. 3. Statistics—Data processing. I. Galwey, Nick. II. Lane, Peter, 1952– . III. Title.
QA276.4.D54 1989 519.5'028'553 88–33023
ISBN 0–19–852201–0
ISBN 0–19–852218–5 (pbk.)

Origination by Computerised Typesetting Services Ltd.,
Redhill, Surrey
Printed in Great Britain by
St. Edmundsbury Press
Bury St Edmunds, Suffolk

Preface

Genstat is a general statistical program designed to help people summarize and analyse information with computers. As well as making it easy for you to carry out standard statistical operations on data, Genstat also provides you with a high-level statistical language that you can use to program non-standard or novel techniques. The program is available on many types of computer, ranging from microcomputers like the IBM PS/2, through mini computers like the MicroVax II, up to mainframe computers like the ICL 2900. If Genstat is not mounted on the computer you want to use, then contact the distributors of Genstat:

Numerical Algorithms Group
Wilkinson House
Jordan Hill Road
Oxford OX2 8DR England
Telephone: (0865) 511245
Telex: 83354 NAG UK G

This book follows on from *Genstat 5: an introduction* (Lane, Galwey, and Alvey 1987). That book introduces the central features of the Genstat language: how to control the input and transformation of data, and how to design their graphical and tabular presentation. In addition, it describes how to do linear regression analysis, and how to analyse a range of designed experiments. This book introduces you to the use of Genstat for more of the statistical techniques that the program can carry out directly. It also describes how to write *procedures* in the Genstat command language, enabling you to define new commands in the language to do operations or analyses that are not provided directly.

The first three chapters build on the facilities described in *Genstat 5: an introduction* to fit regression models. Chapter 1 describes how to fit regression models to grouped data, and the possibilities of using the regression approach to analyse data from unbalanced experiments. Chapter 2 introduces the class of *generalized linear models*, which can be fitted in Genstat using the same commands as for linear regression. The analysis of contingency tables with log-linear models is the application of generalized linear models described here, though many other standard models such as logit and probit models can be analysed in the same way. Chapter 3 describes how to fit standard nonlinear curves, relating one variable to another with or without information about grouping.

Chapter 4 describes more detailed analyses of designed experiments than were covered in *Genstat 5: an introduction*. The effects of treatments can be partitioned into contrasts to help identify or estimate important components; analyses of qualitative effects can be adjusted with respect to the effects of quantitative variables, using the method known as *analysis of covariance*.

The next three chapters introduce a range of techniques that are usually known as *multivariate analysis*. Chapter 5 describes the method of principal components analysis for looking for patterns in a set of quantitative measurements on individuals; also, the method of canonical variate analysis, or linear discriminant analysis, for differentiating between groups of individuals. Chapter 6 introduces the analysis of distances between individuals using principal coordinate analysis, and shows you how to rotate configurations of points from multivariate analyses by Procrustes rotation. Chapter 7 describes the techniques of *cluster analysis*, which search for groupings of individuals based on a set of measurements.

In Chapters 8 and 9, statistical techniques are put on the sidelines, while some of the advanced programming features of the Genstat language are described. In Chapter 8 the concepts of structured programming, as implemented in Genstat, are illustrated. In Chapter 9 the powerful concept of a Genstat *procedure* is introduced, to allow the formation of a general Genstat program for a specific type of task in the Genstat language. The formation of libraries of procedures that can be invoked automatically is illustrated.

The last two chapters introduce methods of analysis of time series. Chapter 10 describes the Box–Jenkins methodology of selecting, fitting, checking, and finally forecasting from "ARIMA", or autoregressive integrated moving-average models. As well as modelling a single series, methods for relating an output series to one or more input series are also covered. Chapter 11 shows how to use the facilities in Genstat for carrying out Fourier transformation, to construct estimates of the spectral density of a time series. One such method of *spectral analysis* is programmed into a procedure for general use.

The examples used to illustrate this wide range of statistical techniques are drawn from a correspondingly wide range of application areas: we hope this will demonstrate the wide applicability of the techniques, and also of the Genstat program. There are exercises at the end of each chapter to encourage you to try out the methods yourself, and solutions are provided in an appendix in case you need help.

Throughout this book, we assume that you are familiar with the basic features of Genstat covered by *Genstat 5: an introduction*, and we refer to sections of that book. However, this book does not have to be read chapter by chapter. So if you want to find out about doing principal components analysis with Genstat, you can start at Chapter 5. There are, of course, some general features of the Genstat language that are introduced in only one place: if you find something new to you that is not explained, try looking it up in the index. Though this book introduces all the major statistical areas of Genstat, there are still a great many features that are either not described at all, or else are described for a limited range of application. A full description of the whole system can be found in the *Genstat 5 Reference Manual* (Genstat 5 Committee 1987).

If you are familiar with Genstat you will be aware that you can lay out your programs in the command language in many ways. In this book we have imposed some conventions, in common with those adopted for other Genstat documentation. For example, we give system words in full and in capitals, whereas identifiers of data structures are in lower case starting with one capital, and option settings that are strings are all in lower case. We hope that the conventions will help you to recognize what each item is, but also that you will experiment and develop the style of programming that you find most convenient.

Finally, we would like to acknowledge the help of our colleagues in writing this book. Particular thanks are due to Norman Alvey, John Gower, Trevor Lambert, and Granville Tunnicliffe Wilson, who read early drafts and gave valuable advice. We also thank the members of the Genstat 5 Committee, without whom the program, and hence this book, would not exist.

Harpenden and Cambridge P.G.N.D.
October 1988 N.W.G.
P.W.L.

Contents

1 Regression analysis of grouped data

1.1 Introduction

When a set of observations is made, the variation in one numerical variable can sometimes be related to others: for example, the amount of money that people spend on luxuries may depend on their income and age. In other cases, the variation may be explained by dividing the observations into groups: for example, people's expenditure on luxuries may vary according to the region in which they live. The use of Genstat for both types of data analysis has been described in *Genstat 5: an introduction* (Lane, Galwey, and Alvey 1987).

In the first type of analysis, the relationship between a response or dependent variable Y and several explanatory or independent variables X_1, X_2 ... X_m can be described by a regression model of the form

$$Y = b_0 + b_1X_1 + b_2X_2 + \ldots + b_mX_m$$

The least-squares values of the parameters b_0, b_1 ... b_m of such models can be estimated using MODEL and FIT statements in Genstat, and a succession of models can be fitted using TERMS, ADD, DROP, and SWITCH statements (see *Genstat 5: an introduction*, Chapter 4). For the second type of analysis BLOCKS and TREATMENTS statements are used to indicate how the observations are grouped, which must be according to a balanced experimental design, and an ANOVA statement is used to indicate the response variable on which an analysis of variance is to be conducted (see *Genstat 5: an introduction*, Chapter 7). But many sets of data are a mixture of these two types: the response variable can be related to others, but the observations can also be classified into groups, and it cannot be assumed that the same parameter values are appropriate in each group. In Genstat, a factor is used to indicate the group to which each observation belongs, and in this chapter we shall show how factors as well as variates can be used in the regression directives to define a model.

The use of factors in the FIT directive is similar to their use in the TREATMENTS directive. In fact, any dataset that can be analysed by the TREATMENTS and ANOVA directives can also be analysed by the MODEL and FIT directives. For balanced designs the output of the ANOVA directive is the more informative, but for unbalanced data, or for designs whose balance cannot be recognized by the TREATMENTS directive, the FIT directive is a useful alternative. We shall show

how the regression directives can be used for this purpose. We shall also discuss how multi-dimensional regression models can be represented graphically on two-dimensional paper.

1.2 Separate regressions on each group

In a survey on the relationship between expenditure on tobacco and income among farm families in the United States, the country was divided into North-East, North-Central, Southern, and Western regions. The families in each region were divided into classes on the basis of their annual income, and the mean annual expenditure on tobacco in each class was estimated. The data, adapted from *Consumer Expenditure Survey Report* (USDA, 1966), were in a haphazard order with some classes not present in some regions. The information was prepared in a form suitable for reading by Genstat, with region, approximate mid-value of income-class ($), and mean expenditure on tobacco ($) in parallel columns. The first and last few lines of the dataset are shown below:

```
W     1500    40.01
S     1500    51.99
NE    3500    82.39
W     2500    35.79
NE    6750   104.99
.      .       .
.      .       .
.      .       .
S     4500    80.02
NC    1500    43.70 :
```

It may be expected that people's expenditure on tobacco will be related to their income, but smoking is a matter of custom and habit as well as spare cash, and the consumption of people in different regions with the same income may be different. Moreover, the relationship between smoking and wealth may be less marked, or may even be negative, in regions where people are more conscious of the health risks.

The simplest way to deal with such data is to analyse the four regions separately. A Genstat program that performs the regression of tobacco expenditure on income in each region, and presents graphs of the results, is shown below, in the form of the "listing" that will appear in the output when the program is executed.

```
 1  UNITS [NVALUES=38]
 2  FACTOR [LABELS=!T(NE,NC,S,W)] Region
 3  OPEN 'Tobacco.dat'; CHANNEL=2
 4  READ [CHANNEL=2] Region,Income,Expend; FREPRESENTATION=labels

 Identifier   Minimum      Mean   Maximum    Values   Missing
     Income       500      6237     16000        38         0
     Expend     35.79     71.13    149.12        38         0
 5  TEXT [VALUES='Annual income ($) in Region 1'] Regntitl[1]
 6  & [VALUES='Annual income ($) in Region 2'] Regntitl[2]
```

```
 7  & [VALUES='Annual income ($) in Region 3'] Regntitl[3]
 8  & [VALUES='Annual income ($) in Region 4'] Regntitl[4]
 9  FOR Reglevel=1...4; Incmtitl=Regntitl[1...4]
10     RESTRICT Income,Expend,Fexpend; Region .EQ. Reglevel
11     MODEL Expend
12     FIT Income
13     RKEEP FITTEDVALUES=Fexpend
14     GRAPH [YTITLE='Ann. expend. on tobacco -$-'; \
15        XTITLE=Incmtitl; YLOWER=0; YUPPER=160; \
16        XLOWER=0; XUPPER=18000; NROWS=25; NCOLUMNS=61] \
17        Fexpend,Expend; Income; METHOD=line,point
18  ENDFOR
```

This program illustrates many of the features of the Genstat language described in detail in *Genstat 5: an introduction*. The UNITS statement gives the default length, 38, for all structures, and the FACTOR statement sets up a structure called Region that will hold the codes in the first column of data. The LABELS option indicates that each value in this column must be either NE, NC, S, or W—any other code would produce an error message. The OPEN statement connects the data file Tobacco.dat to Channel 2, and the READ statement reads the data from this file. The FREPRESENTATION option indicates that the factor values are represented by their labels, rather than being encoded as 1, 2, 3, and 4. The TEXT statements set up structures that will be needed in the GRAPH statement. The statements between FOR and ENDFOR are executed with Reglev taking the values 1 to 4 in turn. Thus on the first pass through the loop the values of Income, Expend, and a new variate Fexpend, are restricted to Region 1. The regression model is fitted, and the RKEEP statement retains the fitted values in Fexpend for subsequent plotting. The options of the GRAPH statement indicate the titles on the y- and x-axes, the upper and lower limits of the y- and x-axes, and the number of columns and rows in the frame to accommodate the graph on the page. The setting Incmtitl of the XTITLE option refers to each of the text structures in turn, in successive passes through the loop, so that the appropriate title is printed for the graph for each region. The settings of the last two options will be used in most future examples: when Genstat is used interactively, they are set to values appropriate for an on-line terminal by default. The axis limits are given explicitly rather than being chosen by Genstat so that the graphs for the four regions will be comparable. The METHOD parameter indicates that the fitted line and the data values are represented by a line plot and a point plot respectively.

The output from the four executions of the FIT and GRAPH statements is as follows:

```
18.................................................................


***** Regression Analysis *****

 Response variate: Expend
     Fitted terms: Constant, Income
```

*** Summary of analysis ***

	d.f.	s.s.	m.s.
Regression	1	2777.	2777.1
Residual	8	3559.	444.9
Total	9	6336.	704.0

Percentage variance accounted for 36.8

* MESSAGE: The following units have high leverage:
34 0.54

*** Estimates of regression coefficients ***

	estimate	s.e.	t
Constant	63.8	11.0	5.79
Income	0.00356	0.00142	2.50

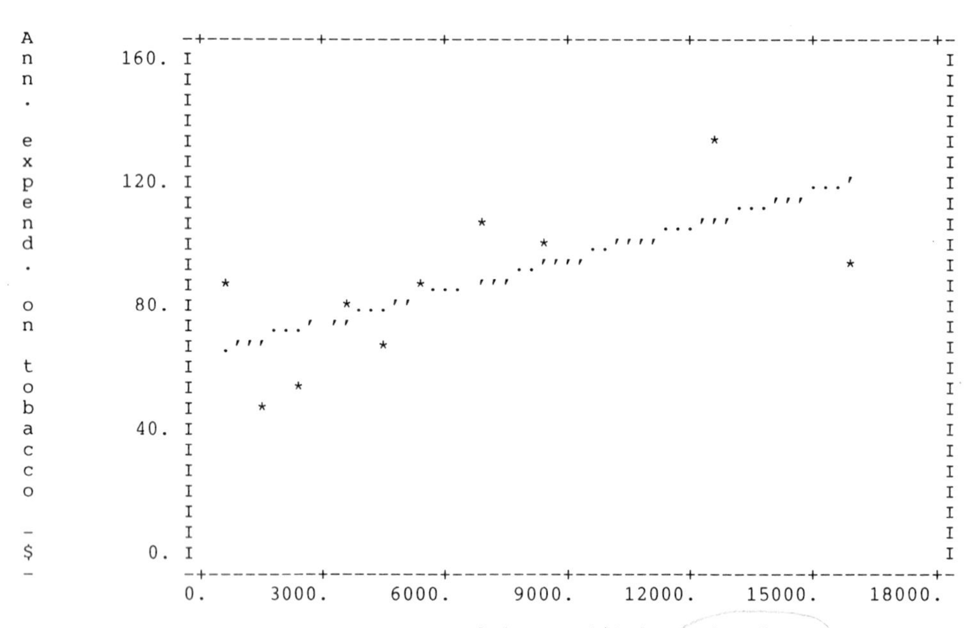

18...

***** Regression Analysis *****

Response variate: Expend
Fitted terms: Constant, Income

*** Summary of analysis ***

	d.f.	s.s.	m.s.
Regression	1	506.	506.1
Residual	7	1262.	180.3
Total	8	1768.	221.1

Percentage variance accounted for 18.4

```
* MESSAGE: The following units have large residuals:
                                  7                2.32

* MESSAGE: The following units have high leverage:
                                  9              0.54


*** Estimates of regression coefficients ***

                   estimate          s.e.            t
Constant              46.33          7.45         6.22
Income             0.001547      0.000924         1.68
```

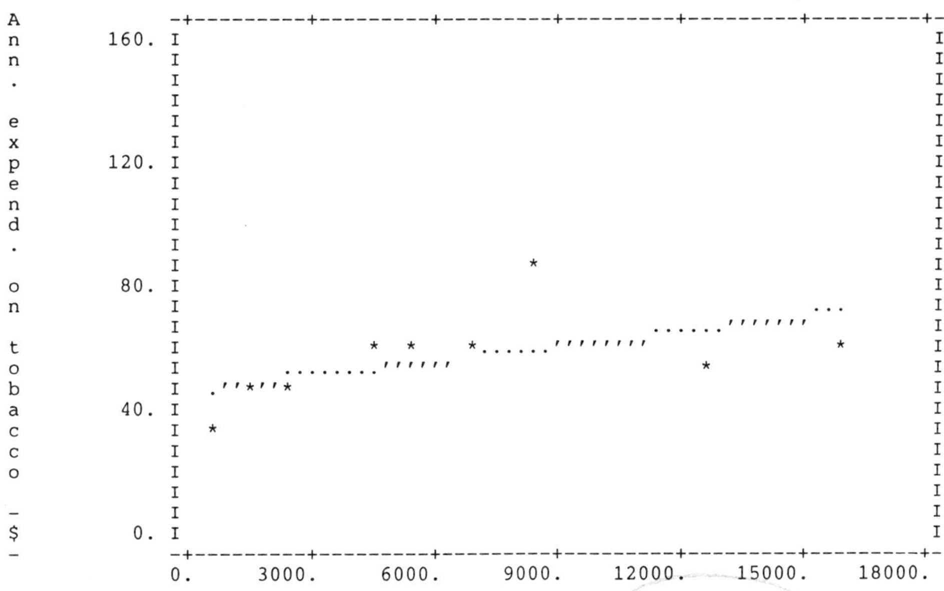

18..

```
***** Regression Analysis *****

 Response variate: Expend
     Fitted terms: Constant, Income


*** Summary of analysis ***

               d.f.          s.s.         m.s.
Regression        1         2007.       2007.3
Residual          7         1006.        143.8
Total             8         3014.        376.7

Percentage variance accounted for 61.8

* MESSAGE: The following units have high leverage:
                                 21              0.55
```

```
*** Estimates of regression coefficients ***

                      estimate        s.e.           t
Constant                57.13         6.44        8.88
Income               0.003030     0.000811        3.74


A             -+---------+---------+---------+---------+---------+---------+-
n        160. I                                                              I
n             I                                                              I
.             I                                                              I
              I                                                              I
e             I                                                              I
x             I                                                              I
p        120. I                                                              I
e             I                                                              I
n             I                                                           ...I
d             I                                                  ....'''     I
.             I                                           ...''*'        *   I
              I              *         *            ...''''                  I
o         80. I                 *        ....''*'                            I
n             I                  ...''''                                     I
              I        .*. ''''                                              I
t             I  ..'''                                                       I
o             I     *                                                        I
b             I                                                              I
a         40. I  *                                                           I
c             I                                                              I
c             I                                                              I
o             I                                                              I
              I                                                              I
_             I                                                              I
$          0. I                                                              I
-             -+---------+---------+---------+---------+---------+---------+-
              0.      3000.     6000.     9000.     12000.    15000.    18000.

                         Annual income ($) in Region 3


18.........................................................................


***** Regression Analysis *****

 Response variate: Expend
     Fitted terms: Constant, Income


*** Summary of analysis ***

               d.f.         s.s.         m.s.
Regression        1        7255.       7255.2
Residual          8        3803.        475.4
Total             9       11059.       1228.7

Percentage variance accounted for 61.3

* MESSAGE: The following units have high leverage:
                               33               0.54


*** Estimates of regression coefficients ***

                      estimate        s.e.           t
Constant                 30.2         11.4        2.65
Income                0.00575      0.00147        3.91
```

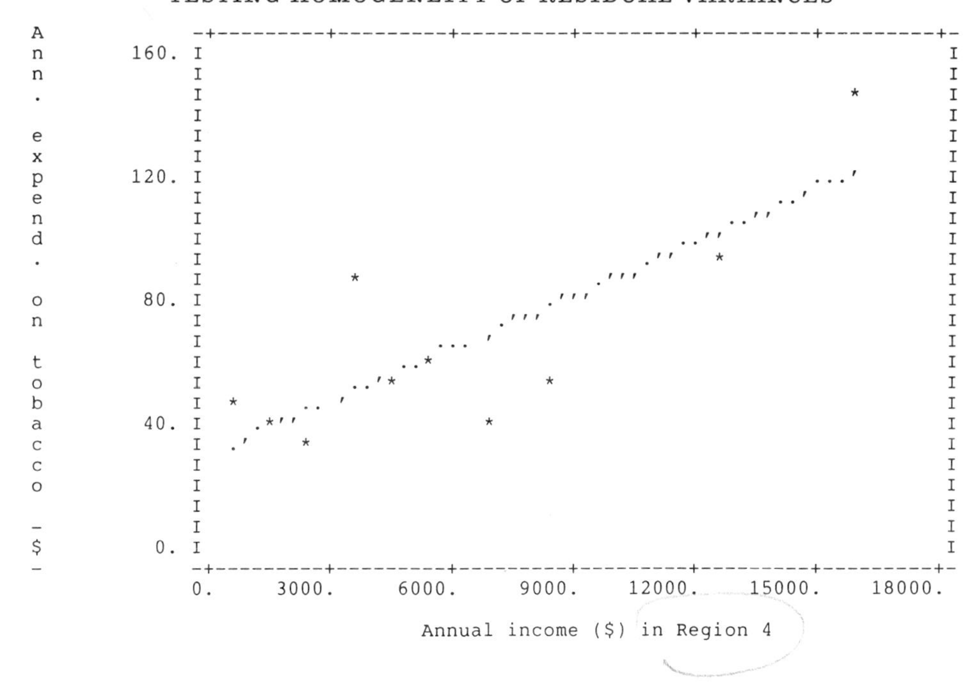

Naturally the intercepts and slopes of the four regression lines are not identical, but this method of analysis provides no evidence as to whether the differences are significant. If they are, explanations can be sought. On the other hand if the four regions have the same regression line, then better estimates of its parameters can be obtained by pooling the data.

1.3 Testing homogeneity of residual variances

In order to test whether the coefficients differ, we need to assume that the residual variances of the four regions are homogeneous. The best known test of this assumption is Bartlett's, which requires the residual sum of squares and residual degrees of freedom of each regression line. After the FIT statement these can be saved by an RKEEP statement, as scalars, using the DEVIANCE and DF parameters respectively, as follows:

```
RKEEP FITTEDVALUES = Fexpend; DEVIANCE = Sdummy; DF = Ddummy
```

The FOR statement is modified to

```
FOR Reglev = 1...4; Incmtitl = Regntitl[1...4]; \
    Ddummy = D[1...4]; Sdummy = S[1...4]
```

so that Ddummy and Sdummy each stand for a different scalar structure on each pass through the loop. Each set of four scalars is then placed in variate of length 4 by the following statements:

```
VARIATE [VALUES = #S[]] ResSS
& [VALUES = #D[]] DF
```

These are placed after the ENDFOR statement. S[] is shorthand for S[1...4]. The hash character (#) has the effect of "unpacking" the structures that follow, so that the values of the scalars, rather than their identifiers, become the values of ResSS and DF. Using the residual sums of squares and degrees of freedom, Bartlett's statistic is obtained as follows:

1. Calculate the average residual variance,

$$s^2 = \frac{\sum_{i=1}^{g} SS_i}{\sum_{i=1}^{g} DF_i}$$

where

g = number of groups (regions),
SS_i = residual sum of squares of the ith group, and
DF_i = residual degrees of freedom of the ith group.

2. Calculate the residual variance of each group,

$$s_i^2 = SS_i/DF_i$$

3. Calculate the approximate test statistic,

$$\chi^2_{\text{approx}} = \left(\log_e (s^2) \times \sum_{i=1}^{g} DF_i\right) - \sum_{i=1}^{g}(DF_i \times \log_e (s_i^2))$$

4. Calculate a correction factor,

$$C = 1 + \left(\frac{1}{3(g-1)}\right) \times \left(\sum_{i=1}^{g} \frac{1}{DF_i} - \frac{1}{\sum_{i=1}^{g} DF_i}\right)$$

5. Adjust the test statistic:

$$\chi^2 = \chi^2_{\text{approx}} / C$$

If the variances are homogeneous, this statistic is distributed approximately as χ^2 with $g-1$ degrees of freedom.

The following Genstat statements calculate and print the test statistic and its degrees of freedom:

```
SCALAR DFchisqd; VALUE = 3
& Logavvar,Smlogvar,Chisqd,Corrfac
CALCULATE Logavvar = LOG(SUM(ResSS) / SUM(DF))
& Resvar = ResSS / DF
& Smlogvar = SUM(DF * LOG(Resvar))
& Chisqd = Logavvar * SUM(DF) - Smlogvar
& Corrfac = SUM(1 / DF) - (1 / SUM(DF))
& Corrfac = 1 + (1 / (3 * DFchisqd)) * Corrfac
& Chisqd = Chisqd / Corrfac
PRINT 'Bartlett''s Test for Homogeneity of Variance' \
& Chisqd,DFchisqd; DECIMALS = 3,0
```

Note that the calculations have been split among more statements than are strictly necessary, for the sake of clarity. The results of a calculation are sometimes placed in one of the structures used in the calculation, to reduce the number of structures declared. In these cases the structure is named according to the values that it will eventually contain. Note also that when a single quote or apostrophe is required in a text structure, it is represented by two single quotes so that it is not interpreted as the end of the text.

In the present example, the PRINT statements produce the following output:

```
Bartlett's Test for Homogeneity of Variance


      Chisqd    DFchisqd
       3.703           3
```

The backslash (\) after the first PRINT statement has the same effect as putting the two statements on one line: it causes both statements to be listed before the output of either is produced, so that the output appears uninterrupted.

A table of the percentage points of χ^2, such as Table 8 in *New Cambridge elementary statistical tables* (Lindley and Scott 1984), shows that a χ^2 value of 3.703 with three degrees of freedom is exceeded by chance over 20% of the time: it is considerably below the 5% critical value of 7.815. A χ^2 value above this value would give us reason to doubt the null hypothesis that the variances are homogeneous, but in the present case the null hypothesis cannot be rejected as unreasonable. However, only the most extreme cases of heterogeneity are reliably detected by this test.

1.4 Fitting a model to the combined data

The simplest way in which the regression models for the regions might differ is by having different intercepts, but the same slope, as shown in Figure 1.1.

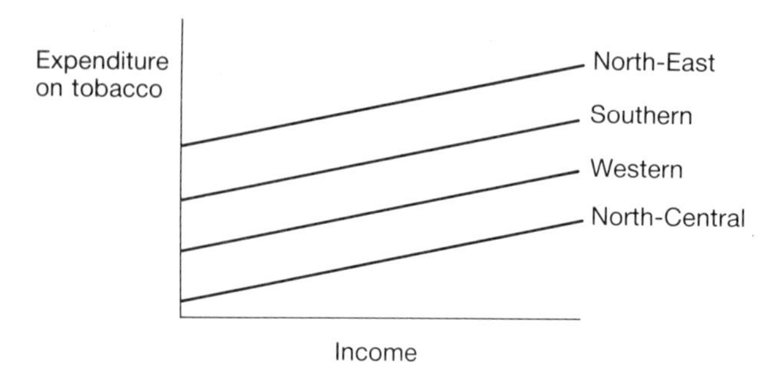

Figure 1.1. Regression model with different intercepts.

We can allow for this situation by extending the regression model

$$\text{Expenditure} = a + b \times \text{Income}$$

to

$$\text{Expenditure} = a_i + b \times \text{Income}$$

where

a_i = the intercept for the ith region.

Alternatively the slope may also vary between regions in which case the model can be further extended to

$$\text{Expenditure} = a_i + b_i \times \text{Income}$$

where

b_i = the slope for the ith region.

In this last model, Region and Income *interact*: the effect of Income (the slope) varies from Region to Region, and likewise the effects of Region (the vertical gaps between the lines) vary according to the level of Income.

In order to modify the Genstat program to fit this new model, the last restriction in force is removed because all the observations are to be considered simultaneously. This is done by the following statement:

```
RESTRICT Income,Expend,Fexpend,Region
```

The interacting factor and variate are indicated in the same way as two interacting factors in a TREATMENT statement (see *Genstat 5: an introduction*, Chapter 7), by the expression Region*Income. The new regression statements are therefore as follows:

```
MODEL Expend
FIT [PRINT = summary,estimates,accumulated] Region * Income
```

The option setting PRINT = accumulated produces an analysis of variance giving the mean square due to adding each term in the model. The other settings ensure that the Summary and Estimates are printed as usual.

The output of the FIT statement is as follows:

```
36..................................................................

***** Regression Analysis *****

*** Summary of analysis ***

               d.f.        s.s.         m.s.
Regression        7      17161.       2451.6
Residual         30       9631.        321.0
Total            37      26793.        724.1

Percentage variance accounted for 55.7

* MESSAGE: The following units have large residuals:
                         17                 2.20
                         33                 2.22
                         34                -2.21

* MESSAGE: The following units have high leverage:
                          9               0.54
                         21               0.55
                         33               0.54
                         34               0.54

*** Accumulated analysis of variance ***

Change                         d.f.        s.s.         m.s.       v.r.
+ Income                          1     10320.3      10320.3      32.15
+ Region                          3      4872.9       1624.3       5.06
+ Income.Region                   3      1968.1        656.0       2.04
Residual                         30      9631.5        321.0

Total                            37     26792.8        724.1

*** Estimates of regression coefficients ***

                            estimate        s.e.          t
Constant                       63.78        9.35       6.82
Income                       0.00356     0.00121       2.94
Region NC                      -17.5        13.6      -1.28
Region S                        -6.7        13.4      -0.50
Region W                       -33.6        13.2      -2.54
Income.Region NC            -0.00201     0.00173      -1.16
Income.Region S             -0.00053     0.00171      -0.31
Income.Region W              0.00219     0.00171       1.28
```

This FIT statement does not fit this model in the precise form in which we gave it above. Instead, it treats the first treatment as a control and compares the slopes and intercepts of the other treatments with it by obtaining the parameters of the model

$$\text{Expenditure} = a_1 + a'_i + (b_1 + b'_i) \times \text{Income}$$

where

a'_i = the difference between the intercepts for the first and ith region,
b'_i = the difference between the slopes for the first and ith region.

Thus the original parameters are related to these new ones as follows:

$$a_2 = a_1 + a'_2 \qquad b_2 = b_1 + b'_2$$
$$a_3 = a_1 + a'_3 \qquad b_3 = b_1 + b'_3$$
$$a_4 = a_1 + a'_4 \qquad b_4 = b_1 + b'_4$$

In the table of regression coefficients, the coefficients for Constant and Income are the same as for the North-Eastern regression line in the previous analysis, since NE is the first level of region.

By calculating

Constant + Region NC = 63.78 − 17.5 ≈ 46.33

we obtain the intercept for the North-Central regression line in the previous analysis, while

Income + Income.Region NC = 0.00356 − 0.00201 ≈ 0.001547

gives its slope. The more natural parameters of the original model can be obtained directly by modifying the FIT statement as follows:

```
FIT [PRINT = estimates; CONSTANT = omit] Region / Income
```

The model Region*Income is expanded to Region + Income + Region.Income, whereas Region/Income is expanded to Region + Region.Income. In the output from the previous FIT statement, the parameter a_1 was a constant which appeared in the model for all regions. The option setting CONSTANT = omit indicates that there is no intercept term, and an additional parameter should be generated by the term Region so that there is one a_i for each region. Similarly the term Income in the previous FIT statement generated a parameter b_1 which was multiplied by income in the model for all regions. When this term is absent, an additional parameter is generated by the term Region.Income so that there is one b_i for each region. The output from the modified FIT statement is as follows:

```
36.....................................................................

***** Regression Analysis *****

*** Estimates of regression coefficients ***

                            estimate         s.e.          t
Region NE                      63.78         9.35       6.82
Region NC                      46.33         9.94       4.66
Region S                       57.13         9.62       5.94
Region W                       30.19         9.35       3.23
Income.Region NE             0.00356      0.00121       2.94
Income.Region NC             0.00155      0.00123       1.26
Income.Region S              0.00303      0.00121       2.50
Income.Region W              0.00575      0.00121       4.75
```

In the output from the original FIT statement, the variance ratios from the summary analysis of variance can be used to determine whether the corresponding terms were worth including. If a term can be omitted from the model, the variance ratio for the term is distributed as $F_{1,30}$, $F_{3,30}$, or $F_{3,30}$ respectively. The values obtained in this example indicate a significant effect of Income on Expenditure, and a significant effect of Region, but no significant interaction. It would therefore be helpful also to know the expenditure of families with an average income in each region. These values can be obtained by the statement:

```
PREDICT [PRINT = se] Region
```

Inserted after the FIT statement, this produces the following output:

```
37..................................................................

*** Predictions from regression model ***

    Table contains predictions followed by standard errors

 Response variate: Expend

      Region
          NE         85.98           5.67
          NC         55.98           5.98
           S         76.02           5.97
           W         66.07           5.67
```

1.5 Graphical presentation of the combined model

Following this combined analysis it is reasonable to plot the four graphs in the same frame. However, joining up the fitted values of successive observations will not produce four straight lines. In order to overcome this problem, we need to have four variates each containing all the fitted values, but each restricted to a different region. The statements to produce these variates, and to plot each of them and the observed expenditures against Income, are as follows:

```
RKEEP FITTEDVALUES = Fexpend
VARIATE Fexpendp[1...4]; VALUES = Fexpend
RESTRICT Fexpendp[1...4]; Region .EQ. 1...4
AXES WINDOW = 1; YTITLE = 'Ann. expend. on tobacco - $ - '; \
    XTITLE = 'Annual income ($)'; YLOWER = 0; YUPPER = 160; \
    XLOWER = 0; XUPPER = 18000
PEN 1...5; METHOD = 4(line),point; LINESTYLE = 1...4,0; \
    SYMBOLS = 4(0),Region; COLOUR = 1
OPEN 'Tobacco.grd'; CHANNEL = 1; FILETYPE = graphics
DGRAPH Fexpendp[1...4],Expend; Income; PEN = 1...5
```

The PEN statement defines five "pens" with different characteristics to be used for plotting the four fitted lines and the data points. The type of line that each pen is to draw is indicated by the METHOD and LINESTYLE parameters: the setting LINESTYLE = 0 indicates that the data points are to be plotted unconnected. The setting of the SYMBOLS parameter indicates that the points plotted are not to be represented by the first four pens, and that the fifth pen is to represent them by the labels of the factor Region. The DGRAPH statement indicates which variables are to be plotted, and associates each plot with the appropriate pen.

The output of this DGRAPH statement is shown in Figure 1.2.

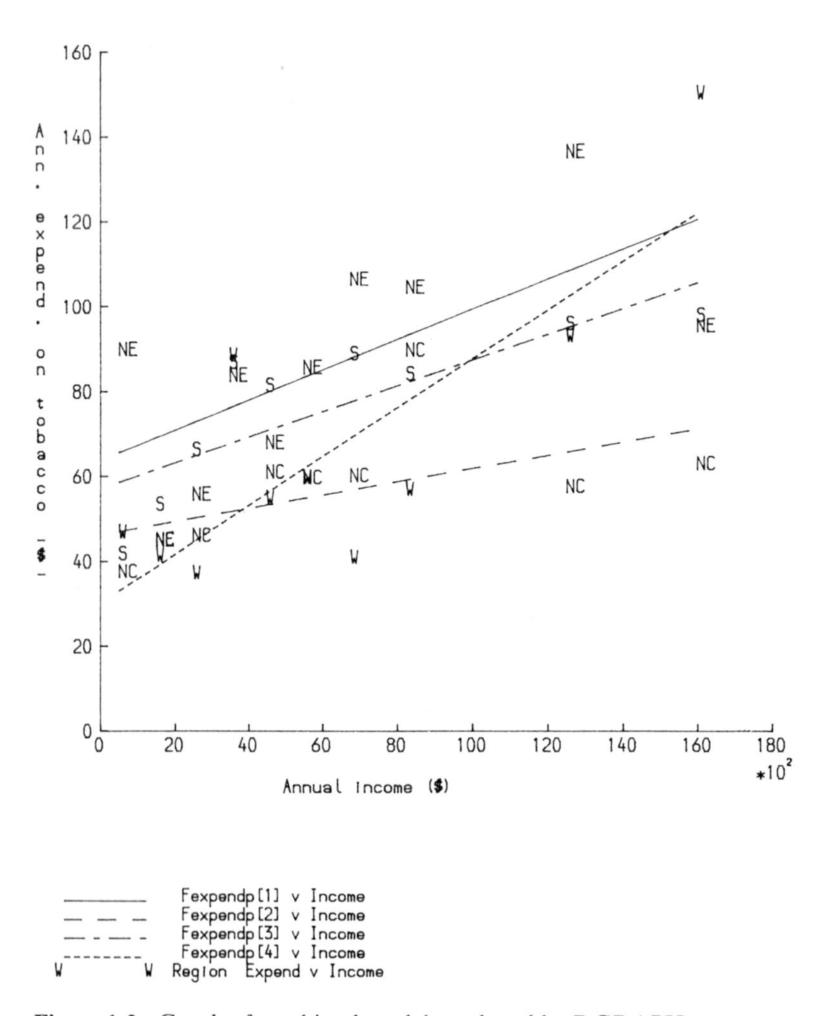

Figure 1.2. Graph of combined model produced by DGRAPH statement.

1.6 The regression approach to analysis of designed experiments

When collecting data for regression analysis it is sometimes possible to control the levels of the explanatory variables so as to maximize the precision of the parameter estimates. For example, if most of the values of the explanatory variable for a regression analysis are clumped in the middle of the range, with only a few at the extremes, the analysis will not give a very precise estimate of the slope of the regression line. On the other hand if all the explanatory values are at the extremes, any curvature in the relationship between the response and explanatory variables is unlikely to be detected. A sensible compromise is often to have the values of the explanatory variable equally spaced over the range, with equal numbers of observations at each value chosen. Such choices often result in a balanced experimental design in which each numerical value of the explanatory variates represents a factor level, and which can be analysed by BLOCKS, TREATMENTS, and ANOVA statements. For example, consider an experiment in which a chemical reaction is studied in the following sets of conditions:

Time (min.)	Temperature (°C)	Time (min.)	Temperature (°C)
80	140	80	150
80	140	80	150
90	140	90	150
90	140	90	150
100	140	100	150
100	140	100	150

Time and temperature can be regarded as two factors with three and two levels respectively, and since each possible combination of these levels occurs the same number of times (here twice), the design is balanced. In some designs, however, the balance depends not only on the combinations of factor levels chosen but also on the numerical values that they represent. Such designs must be analysed using the regression directives.

In an experiment concerning the yield of a chemical reaction at varying temperatures for varying lengths of time, the following data were obtained from *Statistics for experimenters* (Box, Hunter, and Hunter 1978):

Day	Time (min.)	Temperature (°C)	Yield g
1	80	140	78.8
1	90	145	89.7
1	80	150	91.2
1	100	140	84.5
1	90	145	86.8
1	100	150	77.4
2	90	138	81.2
2	90	152	79.5
2	76	145	83.3
2	90	145	87.0
2	90	145	86.0
2	104	145	81.2

The times and temperatures form the pattern shown in Figure 1.3.

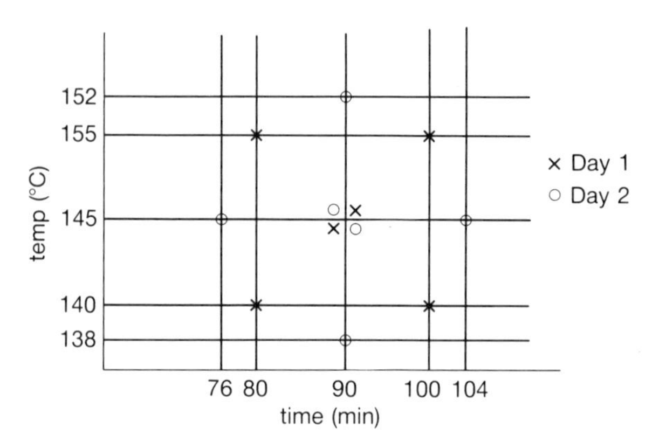

Figure 1.3. Pattern of chemical reaction data.

The following model can be fitted to these data:

$$\text{Yield} = b_1 + b_2 \times D + b_3 \times \text{Time} + b_4 \times \text{Time}^2 + b_5 \times \text{Temperature} + b_6 \times \text{Temperature}^2 + b_7 \times \text{Time} \times \text{Temperature}$$

where

$D = 0$ if Day $= 1$
$D = 1$ if Day $= 2$.

The dummy variable D allows the factor Day to be included in a numerical model. It might be expected that any variation between days would interfere with estimation of the other effects, but if Time and Temperature are regarded as arbitrary axes in two-dimensional space, it can be seen that each day is equally informative about variation in this space. In fact the days are orthogonal blocks, which means that the inclusion of D in the model does not affect the estimates of the parameters b_3 to b_7. This model is fitted to the data by the following Genstat program:

```
UNITS [NVALUES = 12]
FACTOR [LEVELS = 2; VALUES = 6(1,2)] Day
READ Time[1],Temp[1],Yield
 80  140  78.8
 90  145  89.7
 80  150  91.2
100  140  84.5
 90  145  86.8
100  150  77.4
 90  138  81.2
 90  152  79.5
 76  145  83.3
 90  145  87.0
 90  145  86.0
104  145  81.2 :
CALCULATE Time[2],Temp[2] = Time[1],Temp[1] ** 2
& Timetemp = Time[1] * Temp[1]
MODEL Yield
TERMS Day,Time[1,2],Temp[1,2],Timetemp
FIT Day,Time[1,2],Temp[1,2],Timetemp
DROP Day
```

The DROP statement is included to confirm that the parameter estimates are unchanged by omitting DAY from the model. The FIT and DROP statements produce the following output:

```
20..........................................................................

***** Regression Analysis *****

 Response variate: Yield
     Fitted terms: Constant, Day, Time[1], Time[2], Temp[1], Temp[2], Timetemp

*** Summary of analysis ***

              d.f.        s.s.         m.s.
Regression       6      196.96       32.827
Residual         5       15.31        3.063
Total           11      212.28       19.298

Change          -6     -196.96       32.827

Percentage variance accounted for 84.1
```

```
*** Estimates of regression coefficients ***

                   estimate            s.e.            t
Constant             -3989.            645.        -6.19
Day 2                 -1.77            1.01        -1.75
Time[1]               17.89            2.84         6.31
Time[2]            -0.02160         0.00702        -3.08
Temp[1]               45.15            8.29         5.45
Temp[2]             -0.1252          0.0281        -4.46
Timetemp            -0.0975          0.0175        -5.57


  21  DROP Day

21..............................................................

***** Regression Analysis *****

 Response variate: Yield
     Fitted terms: Constant, Time[1], Time[2], Temp[1], Temp[2], Timetemp


*** Summary of analysis ***

              d.f.          s.s.          m.s.
Regression       5        187.56        37.512
Residual         6         24.72         4.120
Total           11        212.28        19.298

Change           1          9.40         9.402

Percentage variance accounted for 78.7


*** Estimates of regression coefficients ***

                   estimate            s.e.            t
Constant             -3978.            748.        -5.32
Time[1]               17.86            3.29         5.43
Time[2]            -0.02147         0.00814        -2.64
Temp[1]               45.00            9.61         4.68
Temp[2]             -0.1247          0.0325        -3.83
Timetemp            -0.0975          0.0203        -4.80
```

The *t*-statistics produced by the FIT statement indicate that all terms are significant, with the exception of Day. The regression coefficients, except for the intercept, are almost unchanged as a result of omitting Day from the model. The slight discrepancies occur because if the time and temperature axes are scaled so that the four outer points indicated by crosses are equidistant from the centre, then the four outer points indicated by circles are not at precisely the same distance from the centre. For example the "Eastern" point deviates from the centre by $104 - 90 = 14$ minutes, and the "North-Eastern" and "South-Eastern" points deviate by $100 - 90 = 10$ minutes. The ratio between these deviations is $14/10 = 1.4$, whereas for perfect orthogonality this ratio, and the corresponding ratios for the other points, should be precisely $\sqrt{2}$ which is 1.4142 correct to four places of decimals.

1.7 Contour plots of response surfaces

Since this model has quadratic terms, the fitted function cannot be plotted simply by joining up the fitted values with straight lines. However, it is possible to save the regression coefficients by placing the following statement after the DROP statement:

```
RKEEP ESTIMATES = Esti
```

Esti is automatically declared as a variate of appropriate length, in this case 6. The coefficients can be placed in separate structures by the following statements:

```
SCALAR B[1...6]
CALCULATE B[1...6] = Esti$[1...6]
```

This CALCULATE statement is equivalent to the six statements

```
CALCULATE B[1] = Esti$[1]
& B[2] = Esti$[2]
```

and so on. The qualifier (for example $[1]) indicates the elements of the structure Esti on which the calculation is to be performed.

The next step is to create variates that will give all combinations of temperature and time over an appropriate range of values, as follows:

```
VARIATE [VALUES = 17(135...155)] Temprang
& [VALUES = (70,72.5...110)21] Timerang
```

These variates can then be used with the coefficients to obtain fitted values of yield at each temperature and time:

```
CALCULATE Vfityld = B[1] + Timerang*(B[2] + B[3]*Timerang) \
  + Temprang*(B[4] + B[5]*Temprang + B[6]*Timerang)
```

The clearest way to present this three-dimensional function graphically on a two-dimensional surface is to plot temperature against time, and to represent the amount of yield by contour lines. This can be achieved by the statements:

```
OPEN 'Reaction.grd'; CHANNEL = 1; FILETYPE = graphics
MATRIX [ROWS = 21; COLUMNS = 17] Fityield; VALUES = Vfityld
AXES WINDOW = 1; YTITLE = 'Temperature'; YLOWER = 135; \
  YUPPER = 155; XTITLE = 'Time'; XLOWER = 70; XUPPER = 110
DCONTOUR [LOWERCUTOFF = 75] Fityield
```

The DCONTOUR statement sends output to the file Reaction.grd, and the contents of this file can be sent to a plotter to obtain a high-resolution contour plot. A low-resolution plot could be obtained with the rest of the output by using the CONTOUR directive, just as the GRAPH directive can be used instead of the DGRAPH directive to produce low-resolution instead of high-resolution graphs.

The DCONTOUR directive needs the data on which to base the contour map to be in a two-way array. A two-way table could be used, but a matrix is more natural here: the MATRIX statement declares the structure Fityield to be a matrix with 21 rows and 17 columns, taking the values (in row order) from the variate Vfityld. Thus the rows correspond to the 21 temperatures 135...155, which will form the y-axis of the picture, and the columns correspond to the 17 times 70,72.5...110, which will form the x-axis.

An option of DCONTOUR has been set to improve the appearance of the picture. The setting LOWERCUTOFF = 75 suppresses all contour lines corresponding to

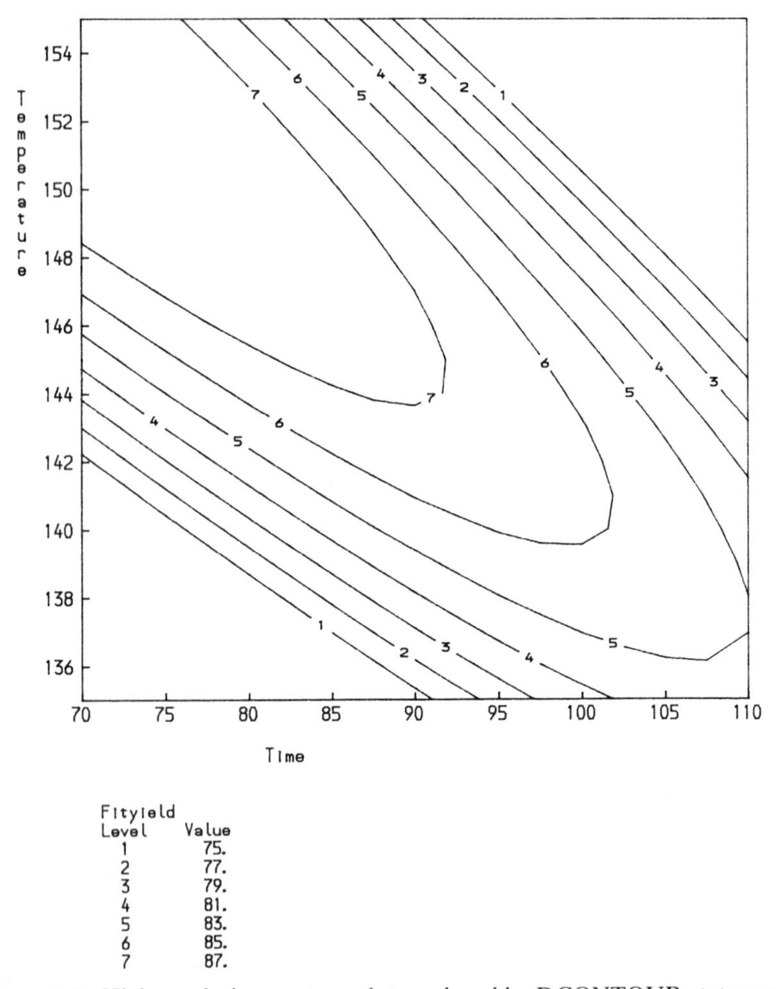

Figure 1.4. High-resolution contour plot produced by DCONTOUR statement.

yields less than 75g, to concentrate attention on the range of values actually achieved in the experiment. The AXIS directive works for the DCONTOUR directive as for the DGRAPH directive, supplying titles for the axes, and specifying the lower and upper bounds for the axis values. The plot obtained from the file Reaction.grd is shown in Figure 1.4.

This plot suggests that the maximum yield is for high temperatures and low times—a combination not actually tested in the experiment. A further experiment could be designed to test such values, if they are actually obtainable in practice.

1.8 Exercises

1(1) In a study of the Andean grain crop quinoa (*Chenopodium quinoa*), two varieties, Baer and Amarilla de Marangani, were grown at a range of sowing densities and of widths between rows of plants, in a randomized block design. The number of days from sowing to maturity and the grain yield were noted. The results obtained are given below. (Unpublished data from J. Risi C.)

Block	Variety	Row width (m)	Sowing density (kg/ha)	Days to maturity	Yield (g/m^2)
1	Baer	0.4	15	148	468
1	Baer	0.4	20	150	511
1	Baer	0.2	15	148	614
1	Baer	0.2	20	148	714
1	Baer	0.2	30	146	641
1	AdeM	0.4	15	240	449
1	AdeM	0.4	20	240	438
1	AdeM	0.2	15	240	541
1	AdeM	0.2	20	238	675
1	AdeM	0.2	30	236	617
2	Baer	0.4	15	152	516
2	Baer	0.4	20	148	499
2	Baer	0.2	15	150	635
2	Baer	0.2	20	148	618
2	Baer	0.2	30	148	562
2	AdeM	0.4	15	238	410
2	AdeM	0.4	20	242	501
2	AdeM	0.2	15	238	494
2	AdeM	0.2	20	238	641
2	AdeM	0.2	30	238	699
3	Baer	0.4	15	150	498
3	Baer	0.4	20	146	491
3	Baer	0.2	15	148	599
3	Baer	0.2	20	148	757
3	Baer	0.2	30	146	661
3	AdeM	0.4	15	242	493
3	AdeM	0.4	20	242	436
3	AdeM	0.2	15	238	511
3	AdeM	0.2	20	236	592
3	AdeM	0.2	30	236	623

Fit the regression of yield on days to maturity separately for each variety. For each regression, obtain a low-resolution graph of the fitted line and the data points. In your GRAPH statement, use the same settings of the YUPPER and YLOWER options for the two varieties, but different settings of XUPPER and XLOWER, as Baer and Amarilla de Marangani have very different ranges of maturity dates.

Conduct a joint regression to determine whether the differences between the varieties in intercept and slope are significant.

1(2) Using the data in Exercise 1(1), fit regression models relating the response variables days to maturity and yield to the explanatory factors block and variety, the explanatory variables row width, sowing density, row width × sowing density, and sowing density^2 and the interaction between variety and the explanatory variables. Note that interactions between variates cannot be included in model formulae. Thus if the variates X[1] and X[2] both interact with the factor F, this is represented in the regression model by

```
F * (X[1] + X[2])
```

Drop the block term from the model, and using all the other terms fit curves relating each response variable to sowing density for Baer in 0.4 m rows, Baer in 0.2 m rows, Amarilla de Marangani in 0.4 m rows, and Amarilla de Marangani in 0.2 m rows. Produce high-resolution graphs of the four curves and the data points in the same frame.

Why is it not possible to perform equivalent analyses using the BLOCKS, TREATMENTS, and ANOVA directives? Show that if row width and its interaction with variety are dropped from the model, the estimates of the parameters of terms involving sowing density are altered.

1(3) An experiment was conducted to investigate the efficiency of the chemical reaction:

$$A + B + C \rightarrow D + \text{other products}$$

Efficiency was measured by the percentage of A converted into D. Five explanatory variables were studied, namely the amount of solvent (in cc), the amount of C (in mol. per mol. of A) the concentration of C (per cent), the time of reaction (in hours) and the amount of B (in mol. per mol. of A). The results obtained are given below, taken from *The design and analysis of industrial experiments* (Davies 1967).

Amount of solvent	Amount of C	Conc. of C	Time	Amount of B	Yield of D (%)
200	4.0	90	1	3.0	34.3
200	4.0	93	2	3.5	51.6
200	4.5	90	2	3.5	31.2
200	4.5	93	1	3.0	45.1
250	4.0	90	2	3.0	54.1
250	4.0	93	1	3.5	62.4
250	4.5	90	1	3.5	50.2
250	4.5	93	2	3.0	58.6

Fit a regression model relating the yield to all the explanatory variables. Obtain a contour plot of the yield against the amount of solvent and the concentration of C. Explain why these are the two explanatory variables most appropriate for use in a contour plot.

Show that the amount of solvent can be dropped from the model without altering the parameter estimates of the other variables. (In fact, the values have been chosen so that all the explanatory variables are mutually orthogonal.) Show that the term

Amount of solvent × Amount of C

can be added to the model, but that

Amount of solvent × Time

cannot.

2 Association between factors: the analysis of contingency tables

2.1 Introduction

In the previous chapter we considered how one numerical variable can depend upon another, and also upon a factor with a limited number of discrete levels which cannot be placed on a numerical scale. However, it is also possible for one factor to depend upon another or others. For example, haemophilia is much commoner in males than females. Here the factors are health, with levels haemophiliac and normal, and sex, with levels male and female. The association is detected by considering the number of observations (that is, individuals) in each category. In this chapter we consider how the associations between two or more factors can be analysed using Genstat, and how the observed associations can be tested against *a priori* hypotheses about the frequencies in each category.

2.2 Testing the independence of classifying factors

A survey was conducted to assess the relationship between the social class of high school students and the curriculum offered by the school that they attended, namely college preparatory, general, or commercial. This was reported in *Elmtown's youth: the impact of social classes on adolescents* (Hollingshead 1946), and further analysed in *Nonparametric statistics for the behavioural sciences* (Siegel 1956). If the data are appropriately ordered they can be read directly into a table. Margins of totals can then be generated (see *Genstat 5: an introduction*, Chapter 6) and the table can be printed. However, before the margins are generated, the values in the body of the table can be copied into a variate for use later in the program. The output of a program that does this is shown below.

```
   1  UNITS [NVALUES=12]
   2  FACTOR [LABELS=!T('I and II',III,IV,V)] Class
   3  & [LABELS=!T('Coll. Prep.',General,Commercial)] Curriclm
   4  GENERATE Class,Curriclm
   5  TABLE [CLASSIFICATION=Class,Curriclm] Tstudent
   6  READ Tstudent

    Identifier   Minimum      Mean   Maximum    Values   Missing
      Tstudent      1.00     32.50    107.00        12         0
  11  VARIATE Students; VALUES=Tstudent
  12  MARGIN Tstudent
  13  PRINT Tstudent; DECIMALS=0
```

Tstudent Curriclm Class	Coll. Prep.	General	Commercial	Margin
I and II	23	11	1	35
III	40	75	31	146
IV	16	107	60	183
V	2	14	10	26
Margin	81	207	102	390

Although the social classes have been given numerical labels, it is probably unwise to judge how far apart they are on a linear scale, particularly since classes I and II have been merged so as to obtain an adequate number of data in each category. This doubt is reflected in the use of Roman numerals, which must of course be treated as text in Genstat, to label the classes.

In this case there is no *a priori* hypothesis about the numbers of students in each class, or studying each curriculum; that is, about the values in the margins of the table. However, there is a natural hypothesis, which we wish to test, that social class and curriculum are not associated. Like all null hypotheses, this one can be tested by comparing the observed data with what would be expected if it were true.

If it were true, the numbers of students in each cell of the table would be predicted adequately by the following model:

$$N_{ij} = Np_{i.}p_{.j} + e_{ij}$$

where

N_{ij} = the number of students from the ith social class studying the jth curriculum, a Poisson variable with mean $Np_{i.}p_{.j}$,
N = the total number of students in the sample,
$p_{i.}$ = the proportion of students from the ith social class (that is, the number in the ith row divided by N),
$p_{.j}$ = the proportion of students studying the jth curriculum, and
e_{ij} = the residual value, a random variable with mean 0.

If certain social classes tend to study a particular curriculum, then the values of e_{ij} will be suspiciously large (positive or negative). We would then need to add to the model the term r_{ij}, specific to each combination of social class and curriculum:

$$N_{ij} = Np_{i.}p_{.j}r_{ij} + e_{ij}$$

We shall decide whether the e_{ij} lie within the acceptable range of random variation by a method rather different from Siegel's (1956), one that is at least as reliable and is more readily extended to multi-way classifications. Leaving out the residual, the first model can be expressed as

$$\log(N_{ij}) = \log(N) + \log(p_{i.}) + \log(p_{.j})$$

and the second as

$$\log(N_{ij}) = \log(N) + \log(p_{i.}) + \log(p_{.j}) + \log(r_{ij})$$

These are factorial models, with and without an interaction term, of the kind commonly used in analysis of variance, and these, as was seen in the previous chapter, can be treated as linear models for regression analysis. However, rather than assume that the logarithms of the counts are Normally distributed, as is necessary for linear regression, we prefer to assume that the actual counts have a Poisson distribution: this is often referred to as a *log-linear model*. Models of this kind, which are linear after a transformation and assume for the response variable a distribution like the Poisson, are called *generalized linear models*. *Generalized linear models* (McCullagh and Nelder 1983) gives a full description of the whole class of such models; you can use Genstat to fit all the standard models in the class. The model we want can be fitted by the following Genstat statements:

```
MODEL [LINK = log; DISTRIBUTION = poisson] Students
TERMS Class * Curriclm
FIT Class + Curriclm
```

The options of the MODEL statement specify the link or transformation function and the distribution for the response variable, but in every other respect the statements are the same as for a linear regression analysis. Note that any factorial model that would be valid in a BLOCKS or TREATMENTS statement is also valid in a FIT, ADD, or other regression statement, provided that it has been included in the maximal model defined in the TERMS statement. This condition is met in the present example, since Class*Curriclm is shorthand for Class + Curriclm + Class.Curriclm.

These statements produce the following output:

```
16..................................................................


***** Regression Analysis *****

 Response variate: Students
     Distribution: Poisson
    Link function: Log
     Fitted terms: Constant + Class + Curriclm


*** Summary of analysis ***
    Dispersion parameter is 1

              d.f.      deviance     mean deviance
Regression       5        274.36          54.87
Residual         6         65.65          10.94
Total           11        340.01          30.91

Change          -5       -274.36          54.87
```

```
* MESSAGE: The following units have large residuals:
                                1                    5.46
                                2                   -2.91
                                3                   -4.20
                                4                    2.38
                                7                   -6.23
                                9                    2.69


*** Estimates of regression coefficients ***

                                  estimate          s.e.          t
Constant                             1.984         0.196      10.14
Class III                            1.428         0.188       7.60
Class IV                             1.654         0.184       8.98
Class V                             -0.297         0.259      -1.15
Curriclm General                     0.938         0.131       7.16
Curriclm Commercial                  0.231         0.149       1.55
* MESSAGE: s.e.s are based on dispersion parameter with value 1
```

In the summary of the analysis it is the residual deviance which is of most interest. This deviance is defined as $-2 \log_e(l/l_{max})$ where

l = the likelihood of the fitted model, and

l_{max} = the likelihood of a model that fits the data perfectly.

It gives a measure of the variation among the estimates of e_{ij}, and is distributed approximately as χ^2 if the model is adequate. A χ^2 table, such as Table 8 in Lindley and Scott (1984) shows that the probability of obtaining a value larger than the observed deviance of 65.65 with six degrees of freedom is less than 0.0005. Dividing the deviance by the degress of freedom gives the residual mean deviance, which has an expected value of 1 (the dispersion parameter referred to in the output) if the model is adequate. The residual variation is 10.94 times greater than expected, and we conclude that the curriculum followed depends on social class. The large residuals for some individual units reflect the poor fit of the model.

The second model, which includes the interaction term, is fitted by the statement

```
ADD Class.Curriclm
```

This produces the following output:

```
17......................................................................


***** Regression Analysis *****

 Response variate: Students
     Distribution: Poisson
    Link function: Log
     Fitted terms: Constant + Class + Curriclm + Class.Curriclm
```

```
*** Summary of analysis ***
    Dispersion parameter is 1

                  d.f.      deviance     mean deviance
Regression         11         340.0          30.91
Residual            0           0.0              *
Total              11         340.0          30.91

Change              *         -65.6              *


*** Estimates of regression coefficients ***

                                      estimate        s.e.          t
Constant                                 3.135       0.209      15.04
Class III                                0.553       0.262       2.11
Class IV                                -0.363       0.326      -1.11
Class V                                 -2.442       0.737      -3.31
Curriclm General                        -0.738       0.367      -2.01
Curriclm Commercial                      -3.14        1.02      -3.07
Class III .Curriclm General              1.366       0.416       3.29
Class III .Curriclm Commercial
                                          2.88        1.05       2.75
Class IV .Curriclm General               2.638       0.454       5.81
Class IV .Curriclm Commercial
                                          4.46        1.06       4.21
Class V .Curriclm General                2.684       0.840       3.19
Class V .Curriclm Commercial
                                          4.74        1.28       3.70
 * MESSAGE: s.e.s are based on dispersion parameter with value 1
```

The e_{ij} cannot now be estimated and there is no residual deviance. This is because the model

$$N_{ij} = Np_{i.}p_{.j}r_{ij}$$

fits the data perfectly, given an appropriate choice of r_{ij}. This can be seen by considering the estimates of the terms in the model. It was shown in Chapter 1, Section 1.4, that Genstat calculates the effects of factor levels relative to the first level rather than to the mean or total of all levels. In the present example, this policy leads to the following less symmetrical but equivalent model:

$$N_{ij} = N_{11}r_{i.}r_{.j}r_{ij}$$

where

$$r_{i.} = \frac{p_{i.}}{p_{1.}}$$

and

$$r_{.j} = \frac{p_{.j}}{p_{.1}}$$

Thus the number of students in Class III taking the commercial curriculum, N_{23}, is given by:

$$\log(N_{23}) = \log(N_{11}) + \log(r_{2.}) + \log(r_{.3}) + \log(r_{23})$$

The terms in this log-linear model correspond to the estimates for Constant, Class III, Curriclm Commercial, and Class III.Curriclm Commercial respectively. Thus

$$\log(N_{23}) = 3.135 + 0.553 - 3.14 + 2.88 = 3.428$$

$$N_{23} = \exp(3.428) = 30.8$$

which is the observed value, 31, within the limit of accuracy of the calculations.

2.3 Forming a contingency table

The concepts introduced in the analysis of a two-way contingency table can be extended to the analysis of a multi-way table. When each observation is classified by several factors, it may be convenient to store the factor values for each observation in a dataset and to use Genstat to find the number of observations in each category.

This method of coding data was used in a study of the polar Eskimos of Greenland, in which the blood type of 300 individuals was identified on the basis of several factors (unpublished data provided by A.W.F. Edwards). In addition to the familiar ABO factor, whose full range of levels is A_1, A_2, B, A_1B, A_2B, and O, these included the MN factor, with levels M, N, and MN, and the S factor, with levels S and s. The values obtained for the first few individuals were as follows—each line represents an individual:

```
A1 MN s
 O MN s
A1 MN s
 O M  s
```

The output of a program that reads these data and tabulates the number of individuals in each category is presented below:

```
1  UNITS [NVALUES=300]
2  FACTOR [LABELS=!T(A1,A2,B,A1B,A2B,O)] ABO
3  & [LABELS=!T(M,N,MN)] MN
4  & [LABELS=!T(S,s)] S
5  OPEN 'Eskimo.dat'; CHANNEL=2
6  READ [CHANNEL=2] ABO,MN,S; FREPRESENTATION=labels
7  TABULATE [CLASSIFICATION=ABO,MN,S; PRINT=counts]
```

```
                                Count
                S                   S           s
   ABO         MN
    A1          M                   4          27
                N                   0           7
               MN                   0          39
    A2          M                   0           3
                N                   0           0
               MN                   0           3
     B          M                   0           0
                N                   0           1
               MN                   0           3
   A1B          M                   0           1
                N                   0           0
               MN                   0           1
   A2B          M                   0           0
                N                   0           1
               MN                   0           0
     O          M                  29          59
                N                   0          31
               MN                  18          73
```

There is no rule against one of a factor's labels being the same as its identifier, and it is standard practice to use this somewhat confusing nomenclature for blood groups.

2.4 Grouping factor levels

The above table shows that blood groups involving A_2 and B are very rare among these Eskimos—too rare for any conclusions to be drawn about the association of levels A_2, B, A_1B, and A_2B with levels of other factors. It is therefore sensible to pool all levels of ABO except O into a single level, AB, and create a new factor, ABOsmmry, with levels labelled AB and O. This is done by the following statements:

```
FACTOR [LABELS = !T(AB,O)] ABOsmmry
CALCULATE ABOsmmry = NEWLEVELS(ABO; !(4(1),2(2)))
```

The NEWLEVELS function has two arguments. The first, ABO, indicates that it is from this factor that new levels are to be obtained. The second is a variate whose values indicate that the first four levels of ABO are all to become level 1, while the last two levels are to become level 2. The new values are placed in ABOsmmry, the levels of ABO itself being left unchanged.

The statements

```
TABLE [CLASSIFICATION = ABOsmmry,MN,S] Tnoindiv
TABULATE [COUNTS = Tnoindiv; PRINT = counts]
```

now produces the following output:

```
                          Tnoindiv
                    S         S         s
ABOsmmry           MN
      AB            M         4        31
                    N         0         8
                   MN         0        46
       O            M        29        59
                    N         0        32
                   MN        18        73
```

The table Tnoindiv, which has no margins, contains the information required for a test of association between the factors, but its values need to be transferred into a variate in order to be used by regression statements. This is achieved by the statement

```
VARIATE Vnoindiv; VALUES = Tnoindiv
```

Moreover, the existing factors are not suitable for the analysis, since their number of values is the number of individuals, not the number of elements in the table. Modified factors are required with the same levels, but which specify which element of the table each element of Vnoindiv represents. This raises the question of the order in which the values in the table were transferred to the variate, and the rule is the same as when factor values are generated by a GENERATE statement. Since ABOsmmry is the first factor specified in the TABLE statement, all values from the first level of ABOsmmry, labelled AB, come before all values from the second level. Within each level of ABOsmmry, since MN is the second factor specified, all values from the first level of MN, labelled M, come before all values from the second level, and so on. Within each combination of ABOsmmry and MN the number of observations from the first level of S, labelled S, precedes the number from the second level. That is, the levels of the first factor in the list, ABOsmmry, change most slowly, and those of the last factor, S, change fastest. The lengths and values of the factors can be changed appropriately by the following statements:

```
FACTOR [MODIFY = yes; NVALUES = Vnoindiv] ABOsmmry,MN,S
GENERATE ABOsmmry,MN,S
```

The option setting MODIFY = yes allows the attributes of the factors to be changed, and their number of values is then made the same as that of Vnoindiv; that is, 12, or one value for every category in the table. Values are given to the factors according to the rule that governs the order of values in a table.

The statement

```
PRINT ABOsmmry,MN,S,Vnoindiv; DECIMALS = 0
```

allows us to check that each value is associated with the right factor labels. It produces the following output.

```
ABOsmmry           MN          S     Vnoindiv
      AB            M          S            4
      AB            M          s           31
      AB            N          S            0
      AB            N          s            8
      AB           MN          S            0
      AB           MN          s           46
       O            M          S           29
       O            M          s           59
       O            N          S            0
       O            N          s           32
       O           MN          S           18
       O           MN          s           73
```

2.5 A three-way classification

The null hypothesis that the blood group factors are independent—that is, that the probability that an individual has a particular value for one factor does not depend on what values he or she has for other factors—is tested in the following analysis:

```
  16   MODEL [LINK=log; DISTRIBUTION=poisson] Vnoindiv
  17   TERMS ABOsmmry * MN * S
  18   FIT [PRINT=model,summary] ABOsmmry + MN + S


18..............................................................................


***** Regression Analysis *****

 Response variate: Vnoindiv
     Distribution: Poisson
    Link function: Log
     Fitted terms: Constant + ABOsmmry + MN + S


*** Summary of analysis ***
    Dispersion parameter is 1

              d.f.     deviance    mean deviance
Regression       4       257.32          64.329
Residual         7        49.58           7.083
Total           11       306.90          27.900

Change          -4      -257.32          64.329

* MESSAGE: The following units have large residuals:
                                5                -2.97
                                6                 2.94
                                7                 4.17
                                8                -3.02
                                9                -2.45
                               10                 2.79
```

The critical value for χ^2 with seven degrees of freedom at the 0.1% level is approximately 14.07. The observed residual deviance exceeds this value, indicating that there is some association.

The association can be broken down into four components, namely:

1. Association between ABOsmmry and MN, averaged over both levels of S.
2. Association between ABOsmmry and S, averaged over all levels of MN.

3. Association between MN and S, averaged over both levels of ABOsmmry.
4. Association between S and the particular combination of ABOsmmry and MN, after allowing for 2 and 3 (equivalent to the association between MN and the particular combination of ABOsmmry and S after allowing for 1 and 3, and so on).

These correspond to the two-way and three-way interactions in an analysis of variance.

The significance of each of these associations is tested in the following analysis:

```
  19   ADD [PRINT=*] ABOsmmry.MN
  20   & ABOsmmry.S
  21   & [PRINT=accumulated] MN.S

21.....................................................................


***** Regression Analysis *****

*** Accumulated analysis of deviance ***

Change                                                  mean    mean deviance
                            d.f.      deviance      deviance           ratio
+ ABOsmmry
+ MN
+ S                            4       257.317        64.329           32.05
+ ABOsmmry.MN                  2         3.001         1.500            0.75
+ ABOsmmry.S                   1        17.081        17.081            8.51
+ MN.S                         2        25.484        12.742            6.35
Residual                       2         4.014         2.007

Total                         11       306.898        27.900
```

The option PRINT = * in the first ADD statement and, implicitly, in the second, suppresses all printing from these statements, and the option PRINT = accumulated causes the accumulated analysis of deviance from the preceding FIT and ADD statements to be printed. Note that it is not possible to obtain the deviance due to each of several terms in a single statement, as can be done in a linear regression analysis (see *Genstat 5: an Introduction*, Chapter 4). Only the total deviance due to the main effects in the FIT statement is given, and it is necessary to give each two-way interaction term in a separate statement in order to obtain their individual deviances.

The accumulated analysis-of-deviance table shows the deviance explained by adding each two-way interaction term to the model. This analysis is slightly ambiguous, in that the amount of deviance attributed to each term depends slightly on the order in which terms are added to the model. The residual deviance represents the three-way interaction: a good deal more computer time is used, and more output is produced, if this term is added explicitly to the model.

The value 3.001 is less than the critical value of χ^2 with two degrees of freedom at the 5% level, but 17.081 exceeds the 0.1% value with one degree of freedom and 25.484 exceeds the 0.1% value with two degrees of freedom. However, 4.014 is less than the 5% critical value with two degrees of freedom. Thus two of the two-way interactions, but not the three-way interaction, are significant.

In order to interpret these results we need to know which levels of ABOsmmry and MN are associated with which level of S. This can be revealed by using two PREDICT statements which are shown, with their output, below:

```
   22  PREDICT ABOsmmry,S

22.....................................................................

*** Predictions from regression model ***

The predictions have been standardized by averaging
fitted values over the levels of some factors:

          Factor   Weighting policy   Status of weights
              MN   Marginal weights   Constant over levels of other factors

  Response variate: Vnoindiv

               S           S          s
        ABOsmmry
              AB        1.33      28.33
               O       15.67      54.67

   23  & MN,S

23.....................................................................

*** Predictions from regression model ***

The predictions have been standardized by averaging
fitted values over the levels of some factors:

          Factor   Weighting policy   Status of weights
        ABOsmmry   Marginal weights   Constant over levels of other factors

  Response variate: Vnoindiv

               S           S          s
              MN
               M       16.50      45.00
               N        0.00      20.00
              MN        9.00      59.50
```

The total number of AB-S individuals is 4, so that the number predicted on average at each of the three levels of M is 4/3 = 1.33. The preceding comments on weighting policy indicate, respectively, that this average is obtained giving equal weight to values at different levels of MN and without regard to association between MN and the other factors.

The first of these tables shows that level S is proportionally commoner among O individuals than among AB individuals. The second table shows that level S is absent among N individuals. It is known that the three blood group factors are not genetically linked, so these associations are likely to be due to genealogical relationships between the individuals in this small community, where inbreeding cannot be entirely avoided.

2.6 Testing hypotheses about the prevalence of categories

So far we have not attempted to interpret the deviance accounted for by the statement:

```
FIT [PRINT = model,summary] ABOsmmry + MN + S
```

This deviance is due to variation in the total number of individuals at different levels of ABOsmmry, of MN, and of S, and it seems unimportant since there is no reason to presume that the different factor levels will be equally prevalent. However, we can devise hypotheses about the prevalence of factor levels based on the larger population of Eskimos in North America from which the polar Eskimos of Greenland were presumably derived. The following approximate proportions are adapted from *The distribution of the human blood groups and other polymorphisms* (Mourant, Kopeć, and Domaniewska-Sobczak 1976):

ABOsmmry		MN		S	
AB	0.7744	M	0.5625	S	0.3600
O	0.2256	N	0.0625	s	0.6400
		MN	0.3750		

Assuming that there is no association between these factors in the larger population, the proportions can be multiplied to give the predicted proportion of individuals in each category. Thus the predicted proportion of AB-M-S individuals is $0.7744 \times 0.5625 \times 0.3600 = 0.1568$.

These proportions are calculated by the following statements:

```
VARIATE [VALUES = 6(0.7744,0.2256)] ABOprob
& [VALUES = 2(0.5625,0.0625,0.3750)2] MNprob
& [VALUES = (0.36,0.64)6] Sprob
CALCULATE Prob = ABOprob * MNprob * Sprob
```

If these predictions were satisfactory, then the number of individuals in each category would be adequately predicted by the model

$$N_{ijk} = Na_{ijk} + e_{ijk}$$

where

N_{ijk} = the number of individuals in the ijkth category: that is, the ith level of ABOsmmry, the jth level of MN, and the kth level of S,

N = the total number of individuals,
a_{ijk} = the predicted proportion of individuals in the *ijk*th category, and
e_{ijk} = the residual value, distributed as in the previous examples.

The model can be expanded to allow for the possibility that particular levels of ABOsmmry, MN, or S predominate or that particular combinations of these factors predominate, by adding terms as follows:

$$N_{ijk} = Na_{ijk}p_{i..}p_{.j.}p_{..k}r_{ij.}r_{i.k}r_{.jk}r_{ijk} + e_{ijk}$$

Leaving out the residual term we can express this as a linear model, as follows:

$$\log(N_{ijk}) = \log(N) + \log(a_{ijk}) + \log(p_{i..}) + \log(p_{.j.}) + \log(p_{..k}) + \log(r_{ij.}) + \log(r_{i.k}) + \log(r_{.jk}) + \log(r_{ijk})$$

However, the term log a_{ijk} does not have to be estimated: it is a separate variate supplied in the data. It is equivalent to the term b log a_{ijk}, where the parameter b is required to take the value 1. Such a term is said to be an *offset* and can be included in a model by using the OFFSET option of the MODEL directive. However, the offset variable is included with the other terms in their linear form—that is, on the logarithmic scale—and so it is necessary to transform it to natural logarithms before specifying the model. The model

$$N_{ijk} = Na_{ijk} + e_{ijk}$$

can thus be analysed by the following statements:

```
CALCULATE Logprob = LOG(Prob)
MODEL [LINK = log; DISTRIBUTION = poisson; OFFSET = Logprob] \
    Vnoindiv
TERMS ABOsmmry * MN * S
FIT [PRINT = summary]
```

No variate or factor is specified in the FIT statement, since only the constant N is to be evaluated. These statements produce the following output.

```
32......................................................................

***** Regression Analysis *****

*** Summary of analysis ***
    Dispersion parameter is 1

              d.f.     deviance    mean deviance
Regression      0          0.0              *
Residual       11        147.4          13.40
Total          11        147.4          13.40

Change          0          0.0              *
```

```
* MESSAGE: The following units have large residuals:
                              1                -3.16
                              4                 2.61
                              5                -3.07
                              6                 6.19
                              7                -3.09
                              8                -3.35
                              9                -2.31
                             10                 5.90
                             11                -2.74
                             12                 2.44

* MESSAGE: The following units have high leverage:
                              8               0.279
                             12               0.186
```

The residual deviance exceeds the 0.1% critical value of χ^2 with 11 degrees of freedom, indicating that the numbers of individuals in the different categories deviate significantly from those predicted by the above model. In order to estimate the effects of $p_{i..}$, $p_{.j.}$, and $p_{..k}$ and to obtain the accumulated analysis of deviance for all the terms including the constant, the statements

```
ADD [PRINT = *] ABOsmmry
& MN
& [PRINT = accumulated] S
```

are added. Their output is as follows:

```
35.................................................................

***** Regression Analysis *****

*** Accumulated analysis of deviance ***
```

Change	d.f.	deviance	mean deviance	mean deviance ratio
+ ABOsmmry	1	8.124	8.124	1.15
+ MN	2	36.807	18.403	2.60
+ S	1	52.928	52.928	7.47
Residual	7	49.581	7.083	
Total	11	147.439	13.404	

The accumulated analysis of deviance indicates that the effects of ABOsmmry are significant at the 1% level, and those of MN and S at the 0.1% level. The residual deviance is the sum of the deviances obtained in the previous analysis for the two-way interactions and the residual, and if the two-way interactions were added they would each account for the same deviance as before. This is because the predicted proportions embody the hypothesis that the factors do not interact.

If PREDICT statements are used after this analysis, the results given are related to the deviations of the observed numbers from those expected rather than the observed numbers themselves, and are not straightforward to interpret. Instead, therefore, we shall obtain the observed proportions from the table Tnoindiv formed in Section 2.4 by the following statements:

```
TABLE [CLASSIFICATION = ABOsmmry] ABOpobs
& [CLASSIFICATION = MN] MNpobs
& [CLASSIFICATION = S] Spobs
CALCULATE ABOpobs,MNpobs,Spobs = \
    TTOTALS(Tnoindiv) / SUM(Tnoindiv)
PRINT ABOpobs,MNpobs,Spobs; DECIMALS = 4
```

Note that a single expression on the right hand side of the CALCULATE statement produces different results according to the table into which the data are to be put. The table Tnoindiv is classified by ABOsmmry, MN, and S, whereas ABOpobs is classified only by ABOsmmry, and the first time the TTOTALS function is used it therefore obtains, at each level of ABOsmmry, the total over all levels of MN and S. These totals are then divided by the grand total given by the SUM function. The second time the TTOTALS function is used it obtains, at each level of MN, the total over all levels of ABOsmmry and S, and the third time it obtains, at each level of S, the total over all levels of ABOsmmry and MN. The PRINT statement produces the following output:

```
                     ABOpobs
ABOsmmry
        AB            0.2967
         O            0.7033

                      MNpobs
        MN
         M            0.4100
         N            0.1333
        MN            0.4567

                       Spobs
         S
         S            0.1700
         s            0.8300
```

These results show that blood group O is much rarer among the polar Eskimos of Greenland than among other Eskimos, and that blood groups M and S are rather commoner.

2.7 Exercises

2(1) In a survey of farms in Audubon County, Iowa, the farms in each of three soil fertility groups were classified according to tenure. The results obtained were as follows, from *Statistical methods* (Snedecor and Cochran 1980):

		Tenure	
Soil	Owned	Rented	Mixed
I	36	67	49
II	31	60	49
III	58	87	80

Obtain the margins of totals of this table, and carry out an analysis of deviance to determine whether there is an association between soil fertility and tenure.

2(2) Tomato seedlings from the second generation of a cross between two inbred lines were classified according to whether the shape of their leaves was "cut" or "entire", whether the colour of their leaves was green or yellow, and whether their stems were purple or green. The results obtained were as follows:

		Stem colour	
Leaf shape	Leaf colour	Purple	Green
Cut	Green	57	5
	Yellow	15	9
Entire	Green	13	11
	Yellow	1	7

According to a genetical hypothesis, the ratios of the numbers of cut to entire, green to yellow, and purple to green are all 3:1, so that if these factors are independent the types of seedlings are expected to occur in the following ratios:

		Stem colour	
Leaf shape	Leaf colour	Purple	Green
Cut	Green	27	9
	Yellow	9	3
Entire	Green	9	3
	Yellow	3	1

Test the extent to which the observed numbers deviate from these ratios, and the strength of the associations between the factors. Obtain and print the three two-way tables of total seedling numbers.

2(3) In an investigation of age distribution in a population of the great tit (*Parus major*), the following numbers of breeding individuals were obtained (Bulmer and Perrins 1973):

Age (yrs)	1	2	3	4	5	6	7	8	9 and over
Males	83	67	31	31	16	7	3	2	0
Females	92	37	22	12	6	3	0	0	0

The expected proportion of individuals in each category, if the proportion dying at each age is constant, is given below, firstly on the hypothesis that this proportion is the same for males and females (H1), then on the hypothesis that it differs between the sexes (H2).

Age		1	2	3	4	5	6	7	8	9 and over
H1	Males	0.2606	0.1440	0.0796	0.0440	0.0243	0.0134	0.0074	0.0041	0.0051
	Females	0.1868	0.1032	0.0570	0.0315	0.0174	0.0096	0.0053	0.0029	0.0036
H2	Males	0.2357	0.1403	0.0835	0.0497	0.0296	0.0176	0.0105	0.0062	0.0092
	Females	0.2189	0.1041	0.0495	0.0236	0.0112	0.0053	0.0025	0.0012	0.0011

Test the fit of the data to these hypotheses, pooling the values for ages of five or more. Genstat cannot give the correct degrees of freedom for the terms in these models: why not? What are the correct degrees of freedom? How much deviance is explained by allowing the expected proportion dying to vary between the sexes?

3 Nonlinear regression analysis

3.1 Introduction

In the first two chapters we have shown how to fit some types of models of dependency of one variable on other variables. However, there are plenty of models of dependency that cannot be cast in the form of linear or generalized linear regression. Often, such nonlinear models can be modified to force them into the linear mould, but the standard assumptions required for fitting the linear models are then often unjustified. Moreover, the results of fitting a modified model are usually inconvenient to interpret in terms of the original variables under study. Methods for fitting nonlinear models without modification are therefore required.

In this chapter we shall describe how to fit the "asymptotic regression" model with Genstat. This is a common form of decay curve, also known by the alternative names "negative exponential" and "Mitscherlich" model. The equation of the curve, for dependence of the variable Y on the variable X, is:

$$Y = a + b \times r^X$$

The parameter a is an asymptote, and b is the range of possible values of Y for positive values of X: $Y = a + b$ if $X = 0$, and $Y = a$ if X is large. The parameter r is a rate parameter that describes the rate of change of the curve relative to the units in which X and Y are measured.

The equation is not linear in the parameter r, though it is linear in a and b. The same equation is often written in an alternative form:

$$Y = a + b \times \exp(- k \times X)$$

where $k = - \log(r)$. The model can be converted into a linear form if the parameter a is known or can be guessed independently, for the equation can be rewritten:

$$(Y - a) = \exp(\log(b) + \log(r) \times X)$$

This form can be fitted as a generalized linear model, as described in Chapter 2, using a logarithmic link function and estimating an intercept ($\log(b)$) and the slope of X ($\log(r)$). Usually, however, it is not possible to treat the parameter a in a special way.

An alternative expression of the model may also be considered:

$$\log(Y - a) = \log(b) + \log(r) \times X$$

This appears to be a linear model relating the logarithm of Y, adjusted by a, to X, and so could be fitted by ordinary linear regression. However, the standard assumptions of linear regression require that the variance of the response variable be constant; in this expression of the model, the response variable is $\log(Y - a)$ and so the assumptions apply to this rather than to Y. If the model is fitted, perhaps incorporating the effects of some factor on the intercept of the regression, then the resulting estimates will be expressed as differences between values of $\log(b)$ rather than of b. Therefore, it is usually preferable to fit this model without recourse to an artificial transformation.

Nonlinear regression models can be fitted in Genstat by the application of a trial-and-error optimization process. For many standard curves, including the asymptotic regression curve, you need only choose the shape of the curve and the form of distribution of the response variable. Genstat will then estimate the parameters by maximizing the likelihood function: this gives the maximum-likelihood estimates. If you want to fit more complicated nonlinear regression models, or need some curve that is not provided as standard, there is a directive FITNONLINEAR to allow you to specify the model in detail. For general information about nonlinear regression, see *Nonlinear regression modeling* (Ratkowsky 1983).

3.2 Fitting an asymptotic regression model

The growth of sugar beet depends on the amount of phosphorus (P) available in the soil, but the relationship is not a linear one. When a soil is rich in phosphorus, extra phosphorus has little effect on the growth of sugar beet, but on poor soils growth is restricted and the size of the plant is closely related to the amount of phosphorus available.

As part of a long-term rotation experiment (Johnston *et al.*, 1986), sugar beet was grown in 1969 on plots whose levels of soil P were very different because of previous manuring policies. The yields and sugar contents of the plants from each of 16 plots were recorded in tonnes per hectare, and the content of bicarbonate-soluble P in the soil was measured in parts per million (ppm). The latter variable is a measure of the phosphorus available to the plants. The following output shows how these data are read with Genstat and displayed in a simple graph.

```
 1  UNITS [NVALUES=16]
 2  OPEN 'Sugarbeet.dat'; CHANNEL=2
 3  READ [CHANNEL=2; PRINT=data] Beetwt,%sugar,SoilP

  1   7.23 18.5  5.4   7.69 18.0  5.4   24.64 20.1  7.8   26.67 19.8  8.0
  2  39.78 19.5 18.0  44.98 19.3 15.6   41.59 19.7 30.4   44.08 19.8 33.8
  3  48.37 19.4 50.4  44.76 19.0 51.0   49.73 18.6 44.0   51.54 18.5 40.2
  4  47.69 19.0 57.2  45.66 19.4 65.0   50.18 18.6 27.0   47.69 18.7 30.0 :
 4  CALCULATE Sugar = Beetwt * %sugar / 100
 5  GRAPH [NROWS=21; NCOLUMNS=61] Sugar; SoilP
```

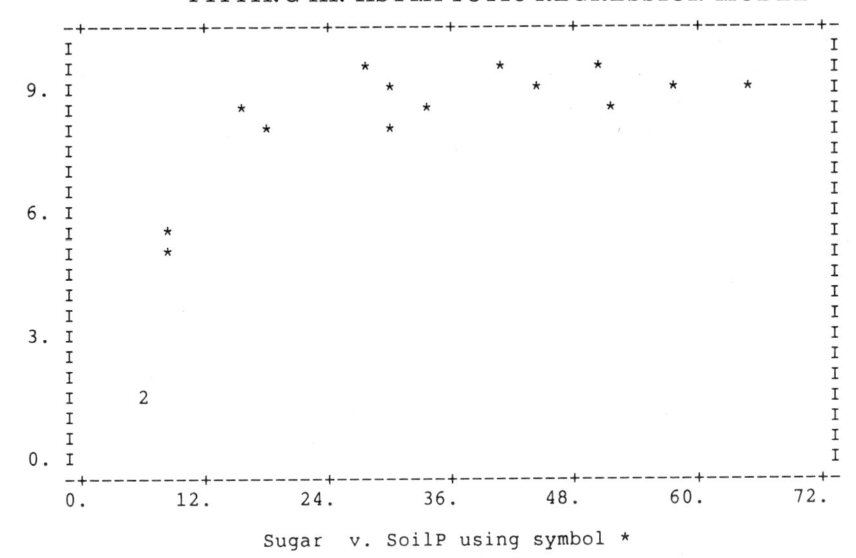

Notice that the lines of data shown by the option PRINT = data in the READ statement are numbered from 1 to 4. When Genstat adds these line numbers to statements or data, it counts lines separately in each channel.

It is clear that the sugar yield is related to the soil phosphorus approximately according to an asymptotic regression model, though there is considerable plot-to-plot variation. (The large variation is at least partly caused by the particularly small size of the plots in this experiment.) The key features of the asymptotic regression curve are that the response variable is constant for large values of the explanatory variable, and it decreases exponentially as the explanatory variable decreases. This model has a theoretical drawback when fitted to positive data, as here, in that it can predict negative values of the response variable for sufficiently small values of the explanatory variable. However, in practice this does not detract from the convenient description provided for the data in the range of the observations actually made.

The asymptotic regression model can be fitted by the directive FITCURVE in Genstat. This directive provides a number of standard nonlinear curves, including other exponential curves such as the double exponential, and growth curves such as the logistic and Gompertz curves. As in linear regression, we must first specify which is the response variate:

```
MODEL Sugar
```

You saw in the last chapter that the MODEL statement has options to specify some attributes of a regression model. For example, the DISTRIBUTION option can be set to specify the distribution to be assumed for the response variable. But we shall assume here the Normal distribution for the yield, as there does not appear to be a noticeable difference in variability of the yields in the range of the data.

The asymptotic regression model is then fitted by a FITCURVE statement:

```
FITCURVE [CURVE = exponential] SoilP
```

The two statements give the following output.

```
   6  MODEL Sugar
   7  FITCURVE [CURVE=exponential] SoilP

7..........................................................................

***** Nonlinear regression analysis *****

 Response variate: Sugar
      Explanatory: SoilP
     Fitted Curve: A + B*R**X
      Constraints: R < 1


*** Summary of analysis ***

              d.f.         s.s.         m.s.
Regression       2      109.076      54.5380
Residual        13        2.536       0.1951
Total           15      111.612       7.4408

Percentage variance accounted for 97.4

* MESSAGE: The following units have large residuals:
                           5                -2.15

* MESSAGE: The following units have high leverage:
                           1              0.49
                           2              0.49
                           3              0.42
                           4              0.44


*** Estimates of parameters ***

                 estimate         s.e.
R                  0.7681       0.0266
B                  -31.31         6.35
A                   8.928        0.135
```

The parameter estimates are given with approximate standard errors: they are approximate because of the nonlinearity of the model, and they should be used with caution. Moreover, the parameter estimates can be highly correlated. This happens particularly when estimates of two parameters are formed effectively from a common subset of the observations. Here, nearly all the information on r and b comes from the four observations with lowest soil phosphorus, whereas the information on a comes mostly from the other 12 observations. The correlations can be displayed for this model in the same way as for any regression model in Genstat: the RDISPLAY directive gives information about the latest model fitted.

```
    8  RDISPLAY [PRINT=correlations]

 8..................................................................

 ***** Nonlinear regression analysis *****

 *** Estimates of parameters ***

                              Correlations
 R                            1.000
 B                            0.978  1.000
 A                            0.344  0.240  1.000
```

In the asymptotic regression model, there is high correlation between the parameters b and r when there are not many observations available for the part of the curve where the slope of the curve changes most. The high correlation is an indication that the two parameters could be changed together, for example increasing b while decreasing r, without much affecting the fit of the model. Thus, the standard error of a single parameter (calculated with the other parameters fixed) may not be a very practical guide to the precision of the estimate.

The summary shows the residual mean square, from which you can derive the standard error of an observation. It is 0.44, which is five percent of the asymptotic yield in this year.

3.3 Trying alternative curves

The curve fitted in the last section appears to fit the data well. However, it may still be of interest to try alternative curves which also have the shape suggested by the graph of the relationship between sugar yield and soil phosphorus. One possibility is to add a linear trend to the relationship; in other words, to hypothesize that yield increases steadily with soil P on soils that are not deficient in P. The equation of this model is:

$$Y = a + b \times r^X + c \times X$$

where c is the rate of steady increase of yield with P.

This curve is also available in FITCURVE. We can fit it by just changing the setting of the CURVE option to 'lexponential', standing for "line plus exponential". Here is the result:

```
    9  FITCURVE [CURVE=lexponential] SoilP

9.........................................................................

***** Nonlinear regression analysis *****

 Response variate: Sugar
      Explanatory: SoilP
     Fitted Curve: A + B*R**X + C*X
      Constraints: R < 1

*** Summary of analysis ***

              d.f.        s.s.        m.s.
Regression       3     109.201     36.4003
Residual        12       2.411      0.2009
Total           15     111.612      7.4408

Percentage variance accounted for 97.3

*** Estimates of parameters ***

                  estimate         s.e.
R                   0.7517       0.0349
B                   -33.92         8.29
C                   0.0082       0.0101
A                    8.589        0.434
```

It is clear that the estimate of the trend parameter c, 0.0082, is small compared to its standard error. Moreover, the residual s.s., 2.411, is only slightly smaller than that for the previous model, 2.536, despite the addition of the extra parameter. Thus we conclude that there is no evidence of an additional trend here.

Another model that might be considered is a logistic or growth-curve model. These curves are S-shaped, and have the additional merit (compared with the exponential curves) that when extrapolated in either direction they tend to some finite quantity rather than to an infinite value. Thus, it might be supposed that a model might result that does not predict negative yields for very small soil P values.

The FITCURVE directive provides three different growth curves. They have equations as follows:

CURVE = logistic $\quad Y = a + c/(1 + \exp(-b \times (X - m)))$
CURVE = glogistic $\quad Y = a + c/(1 + t \times \exp(-b \times (X - m)))^{**}(1/t)$
CURVE = gompertz $\quad Y = a + c \times \exp(-\exp(-b \times (X - m)))$

The logistic curve is symmetrical about the inflexion point, $X = m$, and so is unlikely to be suitable for the data we are modelling. The generalized logistic has four parameters, including the power-law parameter t, and is difficult to fit unless there are data from both ends of the S-shape as well as from the middle portion where the slope is steepest. We shall thus try the Gompertz curve, which has only three parameters and is not constrained to be symmetrical.

Here is the output from fitting the Gompertz curve.

```
  10  FITCURVE [CURVE=gompertz] SoilP

******** Warning (Code OP 1). Statement 1 on Line 10
Command: FITCURVE [CURVE=gompertz] SoilP

Optimization process has not converged
The maximum number of cycles is 20



10..........................................................................


***** Nonlinear regression analysis *****

 Response variate: Sugar
      Explanatory: SoilP
     Fitted Curve: A + C*EXP(-EXP(-B*(X-M)))
      Constraints: B > 0


*** Summary of analysis ***

              d.f.         s.s.         m.s.
Regression       3      109.068      36.3558
Residual        12        2.545       0.2120
Total           15      111.612       7.4408

Percentage variance accounted for 97.2


*** Estimates of parameters ***

                     estimate
B                    0.264933
M                    -7.14471
C                     212.615
A                    -203.685
```

The iterative process of estimating the nonlinear parameters b and m has failed to converge. The default limit on the number of iterations is 20. You can change this by setting the MAXCYCLE option of the RCYCLE directive, as follows:

```
RCYCLE [MAXCYCLE = 50]
```

However, the fact that the process has not converged is an indication that the model is not being fitted successfully. Here is the output from another try, allowing more iterations, and using the 'monitoring' setting of the PRINT option to follow the search for the parameter estimates:

```
 11  RCYCLE [MAXCYCLE=50]
 12  FITCURVE [CURVE=gompertz; PRINT=monitoring] SoilP

12..............................................................................

*** Convergence monitoring ***

Cycle Eval Move      Function value    Current parameters
    0    1    0            51.96138       -1.1190         -1.4032
                              Steps     0.0100000       0.0100000
                              Steps      0.010302       0.0097068
    1   10    0            13.03961      -0.93634         -3.1626
    2   19    0            6.295368       -1.7055         -4.8473
    3   28    0            3.514758       -2.3981         -6.5578
    4   37    0            2.616087       -3.1343         -8.3023
                              Steps      0.011112       0.0089991
    5   46    0            2.572534       -3.5699         -8.7921
    6   55    0            2.555320       -4.3617         -9.3494
    7   64    0            2.548684       -4.5498         -9.6048
    8   73    0            2.547222       -4.6628         -9.7170
                              Steps     0.0094645        0.010566
    9   86    2            2.547133       -4.6694         -9.7257
   10   95    0            2.545918       -4.7824         -9.8333
                              Steps      0.012302       0.0081285
   11  108    2            2.545745       -4.7988         -9.8468
                              Steps     0.0099634        0.010037
   12  122    2            2.545537       -4.8189         -9.8649
   13  131    0            2.545121       -4.8681         -9.9008
                              Steps      0.010827       0.0092364
   14  144    2            2.545081       -4.8700         -9.9074
                              Steps     0.0092388        0.010824
   15  158    2            2.545008       -4.9260        -10.0000
   16  167    0            2.544315       -4.9538         -9.9909
   17  176    4            2.544236       -4.9664        -10.0000
                              Steps      0.011180       0.0089447
   18  204    2            2.924617       -5.9241        -10.0000
   19  213    7            2.570032       -5.2440        -10.0000
   20  222    7            2.544560       -5.0088        -10.0000
   21  231    7            2.544203       -4.9774        -10.0000
                              Steps     0.0075540        0.013238
   22  243    2            2.544189       -4.9773        -10.0000
   23  252    7            2.544201       -4.9770        -10.0000

******** Warning (Code OP 20). Statement 1 on Line 12
Command: FITCURVE [CURVE=gompertz; PRINT=monitoring] SoilP

Fitted curve is close to, or has reached limiting form
No Gompertz fit: limiting case is step function.
```

The monitoring information is given at each iteration of the search process: there are several evaluations of the likelihood function at each iteration, counted by "Eval". Various types of "Move" can be made to try to find a better set of parameter values, and the "Steps" show the magnitude of changes to the parameters being considered. The curve cannot be fitted satisfactorily because there is not enough information about the lower portion of the S-shape. In fact, the best fitting curve is a step function which corresponds to extreme values of the parameters of a Gompertz curve. Genstat gives a warning about this, and will not calculate standard errors, residuals, or leverages. You can look at the best fit that was found, though, because the parameter estimates and fitted values are available.

```
 13  RDISPLAY [PRINT=fittedvalues]

13..............................................................................

***** Nonlinear regression analysis *****

*** Fitted values and residuals ***

                                               Standardized
        Unit  Explanatory      Response  Fitted value  residual  Leverage
           1       5.4000          1.34         1.40         *         *
           2       5.4000          1.38         1.40         *         *
           3       7.8000          4.95         4.92         *         *
           4       8.0000          5.28         5.13         *         *
           5      18.0000          7.76         8.66         *         *
           6      15.6000          8.68         8.42         *         *
           7      30.4000          8.19         8.92         *         *
           8      33.8000          8.73         8.92         *         *
           9      50.4000          9.38         8.93         *         *
          10      51.0000          8.50         8.93         *         *
          11      44.0000          9.25         8.93         *         *
          12      40.2000          9.53         8.93         *         *
          13      57.2000          9.06         8.93         *         *
          14      65.0000          8.86         8.93         *         *
          15      27.0000          9.33         8.90         *         *
          16      30.0000          8.92         8.92         *         *

Mean              30.5750          7.45         7.45         *         *
```

It is clear that the failure is not due to a poor fit, because the fitted values above agree well with the responses. The problem lies in the fact that the parameters of the chosen curve are extreme, and effectively represent a different shape than is typical of that type of curve. Thus, if we really want to fit a growth curve to these data, we shall have to get more information about the response at very low levels of soil phosphorus.

3.4 Fitting curves to grouped data

The plots that gave the data we have been analysing were part of a long-term experiment designed to look at various aspects of crop growth over a number of years. We have found a model that adequately describes the relationship of sugar yield of sugar beet with soil phosphorus in 1969, but we should be wary of assuming this model to be reasonable in all years. Data are in fact available for the three succeeding years, for sugar beet grown on different plots elsewhere in the experiment, according to a crop rotation. The output below shows how the full set of data is read in and a factor is set up to record the year of each observation. The data are plotted to show the form of the relationship between yield and soil phosphorus in each year.

```
1  UNITS [NVALUES=64]
2  OPEN 'Sugar6972.dat'; CHANNEL=2
3  READ [CHANNEL=2] Beetwt,%sugar,SoilP

 Identifier   Minimum       Mean   Maximum     Values   Missing
     Beetwt      1.81      33.29     51.54         64         0
     %sugar     13.50      16.88     20.10         64         0
      SoilP      2.00      28.08     65.00         64         0

4  CALCULATE Sugar = Beetwt * %sugar / 100
5  FACTOR [LEVELS=!(1969...1972); VALUES=16(1969...1972)] Year; DECIMALS=0
6  FOR Level=1969...1972
7     RESTRICT Sugar; Year .EQ. Level
8     GRAPH [NROWS=11; NCOLUMNS=31; YLOWER=0; XLOWER=0] Sugar; SoilP
9  ENDFOR
```

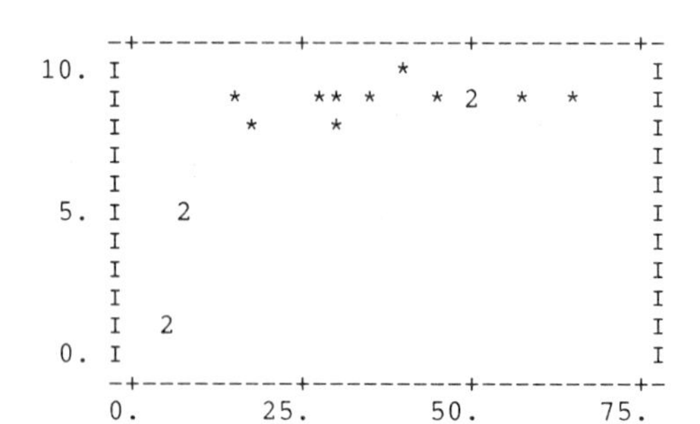

Sugar v. SoilP using symbol *

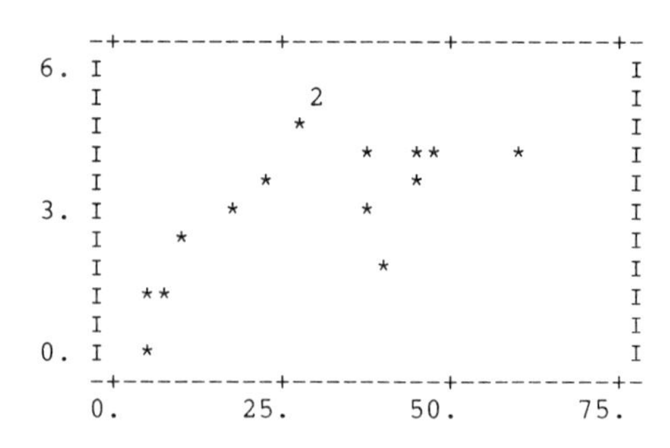

Sugar v. SoilP using symbol *

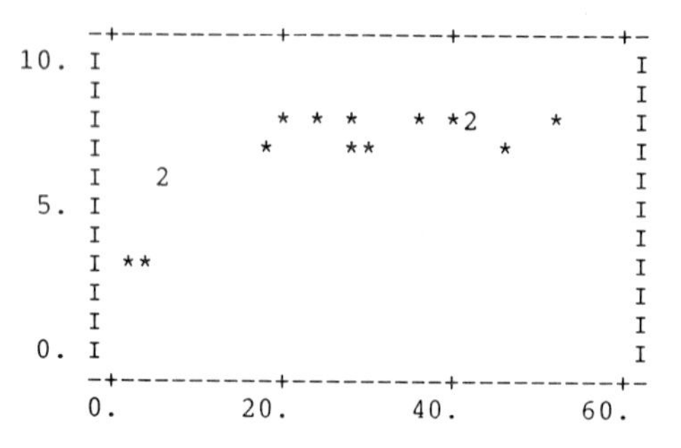

Sugar v. SoilP using symbol *

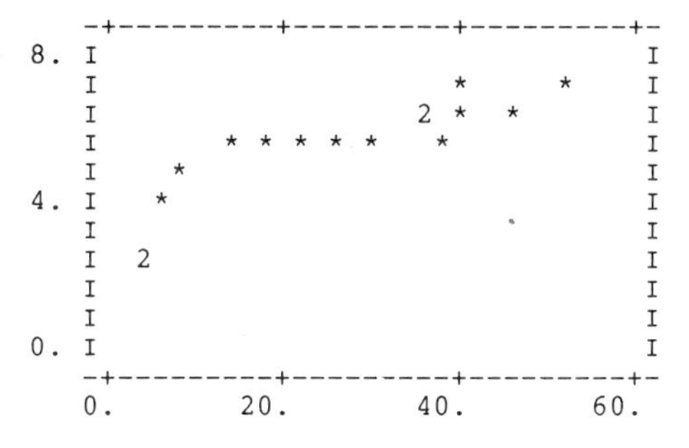

```
          Sugar  v. SoilP using symbol *
```

Before using a model that pools information from each year it is important to check that the observations behave in the same way; that is, that the relationship is of the same form and the variability is similar. This can be done by analysing the data for each year separately in turn, using a RESTRICT statement in a FOR loop:

```
 10   FOR Level=1969...1972
 11      RESTRICT Sugar; Year .EQ. Level
 12      MODEL Sugar
 13      FITCURVE SoilP
 14   ENDFOR

14..........................................................................


***** Nonlinear regression analysis *****

 Response variate: Sugar
      Explanatory: SoilP
     Fitted Curve: A + B*R**X
      Constraints: R < 1


*** Summary of analysis ***

              d.f.         s.s.         m.s.
Regression       2      109.076      54.5380
Residual        13        2.536       0.1951
Total           15      111.612       7.4408

Percentage variance accounted for 97.4

* MESSAGE: The following units have large residuals:
                              5                -2.15

* MESSAGE: The following units have high leverage:
                              1               0.49
                              2               0.49
                              3               0.42
                              4               0.44


*** Estimates of parameters ***

                   estimate        s.e.
R                    0.7681      0.0266
B                    -31.31        6.35
A                     8.928       0.135
```

```
14......................................................................

***** Nonlinear regression analysis *****

 Response variate: Sugar
      Explanatory: SoilP
     Fitted Curve: A + B*R**X
      Constraints: R < 1


*** Summary of analysis ***

              d.f.        s.s.        m.s.
Regression      2        23.13     11.5674
Residual       13        11.85      0.9114
Total          15        34.98      2.3322

Percentage variance accounted for 60.9

* MESSAGE: The following units have large residuals:
                          27               -2.31

* MESSAGE: The following units have high leverage:
                          18               0.61


*** Estimates of parameters ***

                  estimate        s.e.
R                   0.8588      0.0744
B                   -7.72        3.94
A                   4.022       0.336



14......................................................................

***** Nonlinear regression analysis *****

 Response variate: Sugar
      Explanatory: SoilP
     Fitted Curve: A + B*R**X
      Constraints: R < 1


*** Summary of analysis ***

              d.f.        s.s.        m.s.
Regression      2       33.431     16.7155
Residual       13        3.210      0.2469
Total          15       36.641      2.4427

Percentage variance accounted for 89.9

* MESSAGE: The following units have high leverage:
                          34               0.73
                          35               0.42
                          36               0.42


*** Estimates of parameters ***

                  estimate        s.e.
R                   0.8154      0.0409
B                   -6.98        1.13
A                   7.591       0.155
```

```
14..........................................................................

***** Nonlinear regression analysis *****

 Response variate: Sugar
      Explanatory: SoilP
     Fitted Curve: A + B*R**X
      Constraints: R < 1


*** Summary of analysis ***

              d.f.          s.s.          m.s.
Regression       2        30.256       15.1280
Residual        13         2.730        0.2100
Total           15        32.986        2.1990

Percentage variance accounted for 90.5

* MESSAGE: The following units have large residuals:
                             59                 2.05

* MESSAGE: The following units have high leverage:
                             49               0.41
                             50               0.47


*** Estimates of parameters ***

                   estimate         s.e.
R                    0.8666       0.0314
B                     -6.84         1.17
A                     6.377        0.174
```

It seems that the models fit reasonably well in each year, and the residual sums of squares are not very different. The observations are most variable in 1970, but a formal test such as Bartlett's (see Section 1.3) would confirm that the heterogeneity of the variances is not too large to consider pooling the data.

Hence we continue by fitting models to all the data, pooling the information about variability from all the years. First we shall fit a single curve, despite the obvious difference in asymptotic yield in the different years. Fitting this model will give information on the extent of the differences between years.

```
  15  RESTRICT Sugar
  16  MODEL Sugar
  17  FITCURVE [CURVE=exponential] SoilP

17..........................................................................

***** Nonlinear regression analysis *****

 Response variate: Sugar
      Explanatory: SoilP
     Fitted Curve: A + B*R**X
      Constraints: R < 1
```

```
*** Summary of analysis ***

                  d.f.          s.s.          m.s.
Regression           2         149.0        74.490
Residual            61         230.2         3.773
Total               63         379.1         6.018

Percentage variance accounted for 37.3


* MESSAGE: The following units have large residuals:
                                27                -2.55

* MESSAGE: The following units have high leverage:
                                33              0.165
                                34              0.279


*** Estimates of parameters ***

                      estimate          s.e.
R                       0.9025        0.0427
B                        -6.58          1.55
A                        6.908         0.444
```

You can see from the size of the residual mean square that this model does not fit well compared to the separate curves fitted earlier.

The model can easily be modified to allow separate asymptotes for each year. This is done by including the identifier of the grouping factor as well as that of the explanatory variate in the FITCURVE statement. Genstat will then fit a model with separate constant terms, as in linear regression.

```
  18  FITCURVE [CURVE=exponential] SoilP,Year

18.............................................................................


***** Nonlinear regression analysis *****

 Response variate: Sugar
      Explanatory: SoilP
  Grouping factor: Year, constant parameters separate
     Fitted Curve: A + B*R**X
      Constraints: R < 1


*** Summary of analysis ***

                  d.f.          s.s.          m.s.
Regression           5        326.00       65.2004
Residual            58         53.14        0.9162
Total               63        379.14        6.0181

Percentage variance accounted for 84.8

* MESSAGE: The following units have large residuals:
                                 1                -3.42
                                 2                -3.37
```

```
*** Estimates of parameters ***

                                    estimate          s.e.
R                                     0.8853        0.0145
B                                   -7.44228
A   Year 1969                        8.46773
A   Year 1970                        4.22311
A   Year 1971                        8.10381
A   Year 1972                        6.64220
```

The output from this model does not include standard errors for the linear parameters: that is, b and a. Genstat does not attempt to calculate these standard errors because if there were many levels of the grouping factor there would be many parameters, and the calculation of standard errors would require a trial-and-error search in many dimensions.

Note also that the parameterization of the effect of the factor is different from that used by default in linear regression (Section 1.4). This is because it is not possible in nonlinear models to test differences between the levels in the same way as in a linear model, so there is no advantage in expressing the effects of the factor in terms of differences.

The new model is a much better fit than the last, but the residual mean square is still larger than that found with the curves for the individual years.

To try to improve the fit further, a model can be fitted with only the nonlinear parameter, the rate, common to all years. This is done by including the "interaction" of the factor with the explanatory variate, as in a linear model when separate slopes are to be estimated for each level of a factor.

```
  19  FITCURVE [CURVE=exponential] SoilP * Year

 19..........................................................................

***** Nonlinear regression analysis *****

  Response variate: Sugar
       Explanatory: SoilP
   Grouping factor: Year, all linear parameters separate
      Fitted Curve: A + B*R**X
       Constraints: R < 1

*** Summary of analysis ***

               d.f.          s.s.          m.s.
Regression        8        357.58       44.6973
Residual         55         21.56        0.3920
Total            63        379.14        6.0181

Percentage variance accounted for 93.5

* MESSAGE: The following units have large residuals:
                             27                -3.20
                             28                 2.71
```

```
*** Estimates of parameters ***

                                     estimate        s.e.
R                                      0.8120      0.0173
B   Year 1969                        -22.6810
A   Year 1969                         9.01578
B   Year 1970                        -10.3565
A   Year 1970                         3.92967
B   Year 1971                        -7.06415
A   Year 1971                         7.58575
B   Year 1972                        -8.94933
A   Year 1972                         6.25012
```

3.5 Parallel curve analysis

By fitting a series of models to grouped data as in the last section, you can decide what complexity is justified in a model to describe the data adequately. Formally, you can test the differences in the residual mean squares of successive models by F-tests to determine when there is statistical significance in the differences. This process is known as "parallel model analysis"; the information required for it can be produced by Genstat in a more convenient form than you have seen above.

Successive models of the same type can be fitted by giving ADD, DROP, and SWITCH statements, as in linear regression, to specify what changes to make to the set of explanatory variables. An initial FITCURVE statement must be given to define the type of curve, and a TERMS statement must precede it, including all the terms to be used, so that preliminary calculations can be done with all the variables. The models fitted in the previous sections can thus be fitted by the following statements:

```
MODEL Yield
TERMS SoilP * Year
FITCURVE [CURVE = exponential] SoilP
ADD Year
ADD SoilP.Year
```

The advantage of fitting the models in this way is that a summary analysis can then be accumulated over the successive statements. With the FITCURVE statement, the setting 'accumulated' of the PRINT option cannot break down the sum of squares between different explanatory variables in the model; the breakdown can be achieved only by fitting each model explicitly, specifying the changes to construct each model from the previous model. Thus, on giving the statements above with the last one modified to print the accumulated summary:

```
ADD [PRINT = accumulated] SoilP.Year
```

The output from this last statement is as follows.

```
***** Nonlinear regression analysis *****

*** Accumulated analysis of variance ***

Change                          d.f.        s.s.        m.s.       v.r.
+ SoilP                            2    148.9805     74.4903     190.01
+ Year                             3    177.0212     59.0071     150.52
+ SoilP.Year                       3     31.5766     10.5255      26.85
Residual                          55     21.5618      0.3920

Total                             63    379.1402      6.0181
```

The analysis has components for "displacement" (separate asymptotes), represented by the term Year, and for the effect of allowing separate range parameters, represented by the interaction term SoilP.Year.

There is one further step possible in the sequence of models fitted by FITCURVE, and that is to fit separate nonlinear as well as separate linear parameters for each year. We effectively fitted this model in the FOR loop in the last section, but in order to do so while still pooling the information about variability from each curve, we give a further statement with the option NONLINEAR = separate. The output from this statement is given below.

```
  24  ADD [PRINT=model,estimates,accumulated; NONLINEAR=separate]


24.............................................................................


***** Nonlinear regression analysis *****

 Response variate: Sugar
      Explanatory: SoilP
  Grouping factor: Year, all parameters separate
     Fitted Curve: A + B*R**X
      Constraints: R < 1


*** Accumulated analysis of variance ***

Change                          d.f.        s.s.        m.s.       v.r.
+ SoilP                            2    148.9805     74.4903     190.59
+ Year                             3    177.0212     59.0071     150.97
+ SoilP.Year                       3     31.5766     10.5255      26.93
+ Separate nonlinear               3      1.2380      0.4127       1.06
Residual                          52     20.3239      0.3908

Total                             63    379.1402      6.0181


*** Estimates of parameters ***

                              estimate        s.e.
R  Year 1969                    0.7681      0.0376
B  Year 1969                    -31.31        8.98
A  Year 1969                     8.928       0.191
R  Year 1970                    0.8588      0.0487
B  Year 1970                     -7.72        2.58
A  Year 1970                     4.022       0.220
R  Year 1971                    0.8154      0.0515
B  Year 1971                     -6.98        1.43
A  Year 1971                     7.591       0.195
R  Year 1972                    0.8666      0.0429
B  Year 1972                     -6.84        1.59
A  Year 1972                     6.377       0.237
```

It is clear that there is no evidence for different rates in each year, but there is evidence of different ranges and of different asymptotes in each year.

Finally, we use the directives for high-resolution graphics to draw a picture of the fitted curves. The statements below make use of suffixed identifiers such as Fitted[1969], introduced in Chapter 5 of *Genstat 5: an introduction*. They also involve qualified identifiers such as F$[Pos[]], described in Section 1.8 of this book. The RESTRICT statement is described in Chapter 8 of *Genstat 5: an introduction*; here, the SAVE parameter of RESTRICT is set to store in variates the sets of unit numbers defined by the restrictions.

```
RKEEP FITTEDVALUES = F
RESTRICT (Sugar)4; Year.EQ.1969...1972; SAVE = Pos[1969...1972]
RESTRICT Sugar
CALCULATE Fitted[1969...1972],Observed[1969...1972], \
    P[1969...1972] = F$[Pos[]],Sugar$[Pos[]],SoilP$[Pos[]]
OPEN 'Sugar.grd'; CHANNEL = 1; FILETYPE = graphics
AXES 1; YTITLE = 'Sugar yield, t/ha'; XTITLE = 'Soluble P, ppm'
PEN 1...5; LINESTYLE = 4(0),1; METHOD = 4(point),monotonic; \
    COLOUR = 1; SYMBOLS = '9','0','1','2',0
DGRAPH [TITLE = 'Sugar beet, 1969 - 1972'; KEYWINDOW = 0] \
    Observed[],Fitted[]; P[]; PEN = 1...4,4(5)
STOP
```

The KEYWINDOW option is set to 0 in the DGRAPH statement to suppress the key that accompanies the graph by default: it is not needed here because the points are identified by the last digit of the year, as set by the SYMBOLS parameter in the PEN statement.

The picture produced by these statements is shown in Figure 3.1.

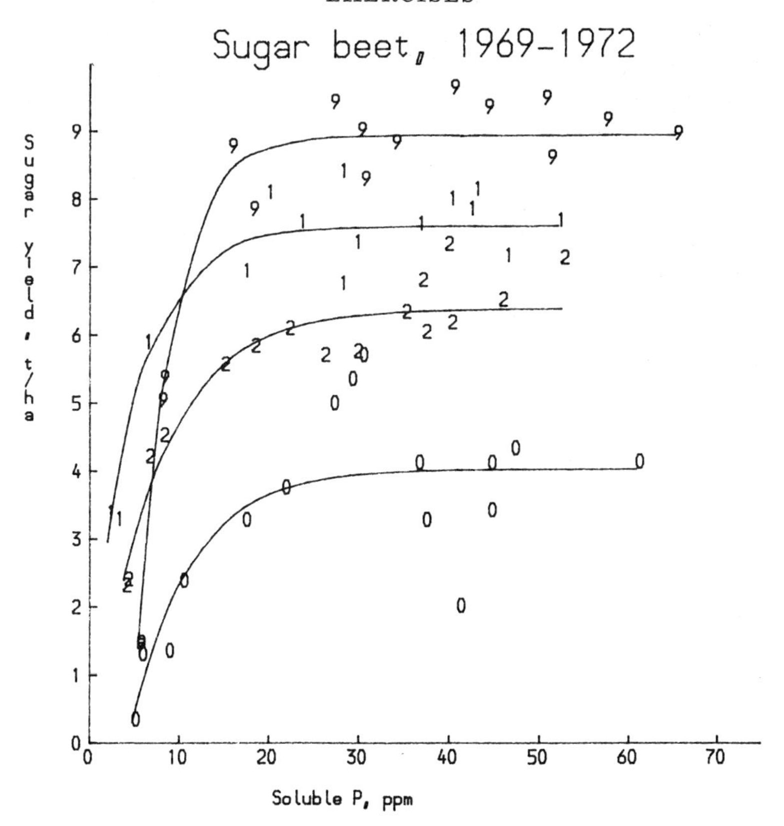

Figure 3.1. High-resolution picture of fitted curves.

3.6 Exercises

3(1) The following measurements of a property of latex were taken at bi-weekly intervals from samples of the material: the first measurement was taken in Week 1, the second in Week 3, and so on.

0.776 0.852 0.850 0.869 0.939 0.904 0.930 0.948 0.942 0.938
0.979 0.975 0.955 0.993 0.985 1.013

(Example from Villars, 1947.)
Fit an exponential curve to the measurements.

3(2) The water content of bean root cells varies according to the distance of the cells from the tip of the root. Fit a logistic curve to the following observations of water content:

Water content	1.3	1.3	1.9	3.4	5.3	7.1	10.6	16.0
Distance	0.5	1.5	2.5	3.5	4.5	5.5	6.5	7.5
Water content	16.4	18.3	20.9	20.5	21.3	21.2	20.9	
Distance	8.5	9.5	10.5	11.5	12.5	13.5	14.5	

(Example from Heyes and Brown, 1956.)

Check the fit of the curve with a plot of the observed and fitted values, and also of the residuals against the fitted values. Try fitting the curve with a weight variable inversely proportional to the response variable, to compensate approximately for the greater variability of water content further from the root tip. This can be done by forming a variate with values equal to the reciprocal of the values of the response variate, and giving the identifier of the weights variate in the setting of the WEIGHTS option of the MODEL statement.

Compare the fitted logistic with a generalized logistic curve, also fitted with weights.

3(3) An experiment was carried out in 1945 and 1946 to relate the yield of potatoes to the amount of phosphate fertilizer used. The experiment tested five levels of the fertilizer in four randomized blocks; the treatment means were as follows:

	Phosphate (lb/acre)				
Year	0	40	80	120	160
1945	232.65	369.08	455.63	491.45	511.50
1946	104.75	188.63	211.75	217.63	231.13

(Example from Gomes, 1953.)

Fit exponential models for the relationship of yield to amount of fertilizer, analysing parallelism of the curves for the two years.

4 The analysis of balanced experimental designs

4.1 Introduction

Experimenters and statisticians have devoted much effort to devising experimental designs that make the data collected as informative as possible. Such efficiency is achieved through balance—for example, by ensuring that the levels of two treatment factors occur in all combinations, or that all treatments occur once in each randomized block. Three of the strengths of Genstat are to allow such balanced experimental designs to be specified simply and succinctly in BLOCKS and TREATMENTS statements, to check that the design is in fact balanced, and to provide a full, and fully appropriate, analysis (see *Genstat 5: an introduction*, Chapter 7).

In this chapter we consider some of the techniques used to make experiments informative, and show how the appropriate analyses can be specified in Genstat.

4.2 Relationships between treatment factors: nesting and crossing

In an experiment to test the uniformity of a manufacturing process, the strength of tyre cord was tested using eight bobbins of cord from each of two plants. There is no sense in which Bobbin 1 from Plant I corresponds to Bobbin 1 from Plant II and hence these two treatment factors cannot be regarded as interacting: any variation between bobbins is *within* plants. *The analysis of variance* (Sheffé 1959) stressed this point by numbering the bobbins from 1 to 16; but numbering them from 1 to 8 within each plant facilitates tabulation of the data using Genstat. The strength of the cord was tested by making two adjacent breaks, which gave duplicate measurements, near the beginning of the cord on each bobbin, and again at 500 yd intervals along the cord, up to 2500 yd. The breaking force was measured in pounds.

The output of a Genstat program that reads and tabulates these data is presented below. The GENERATE statement gives values to the factors and hence indicates the order in which the data were arranged: that is, the values from each plant grouped together in ascending order, within each plant the values from each bobbin grouped together in ascending order, and so on. The factors are given in a different order in the TABLE statement, so as to give a compact table for printing.

```
  1   UNITS [NVALUES=192]
  2   FACTOR [LABELS=!T(I,II)] Plant
  3   & [LEVELS=8; LABELS=*] Bobbin
  4   & [LEVELS=!(0,5...25)] Distance
  5   & [LEVELS=2] Break
  6   GENERATE Plant,Bobbin,Distance,Break
  7   OPEN 'Cord.dat'; CHANNEL=2
  8   READ [CHANNEL=2] Force

 Identifier    Minimum       Mean    Maximum     Values   Missing
      Force      20.30      22.03      23.70        192         0
  9   TABULATE [CLASSIFICATION=Plant,Distance,Break,Bobbin] Force; \
 10      TOTALS=Tforce
 11   PRINT Tforce; FIELDWIDTH=5; DECIMALS=1

                                          Tforce
                                Bobbin       1    2    3    4    5    6    7    8
       Plant        Distance     Break
           I            0.00         1    21.4 21.6 21.7 22.1 21.4 21.4 20.6 21.5
                                     2    21.0 22.5 21.2 22.5 20.7 20.5 21.3 21.7
                        5.00         1    21.3 21.6 22.0 21.6 22.0 20.7 22.0 21.0
                                     2    20.7 21.7 21.0 22.0 20.5 20.7 21.3 21.3
                       10.00         1    21.3 21.7 21.6 21.5 21.6 21.3 22.2 22.0
                                     2    21.8 21.7 21.4 22.0 21.0 21.7 21.3 21.8
                       15.00         1    21.2 22.5 20.9 21.3 21.6 21.5 21.3 22.5
                                     2    21.1 21.1 21.6 21.3 21.1 21.2 21.3 21.4
                       20.00         1    21.5 21.1 21.7 21.6 22.0 20.7 21.4 21.9
                                     2    21.4 21.8 22.0 21.6 21.0 21.4 21.7 21.6
                       25.00         1    20.3 21.9 22.2 22.0 21.8 21.3 22.5 22.2
                                     2    21.9 22.3 22.0 22.4 22.1 21.1 22.0 21.4
          II            0.00         1    22.5 22.4 21.5 22.0 21.4 22.2 21.0 22.5
                                     2    22.3 22.7 22.3 22.4 21.4 23.1 21.6 22.4
                        5.00         1    21.0 22.1 22.7 21.7 22.6 23.0 21.3 22.5
                                     2    22.1 23.0 22.1 23.1 23.4 22.6 22.5 23.0
                       10.00         1    21.7 23.0 21.7 23.0 22.7 22.7 22.7 22.4
                                     2    22.8 22.7 21.5 22.0 22.5 22.7 23.0 23.1
                       15.00         1    22.2 23.3 22.0 23.6 21.6 22.3 21.7 22.7
                                     2    23.0 23.1 21.9 23.3 23.5 22.7 22.8 22.6
                       20.00         1    23.2 22.8 23.3 23.0 22.8 23.7 22.5 23.3
                                     2    22.9 22.5 22.3 22.6 22.4 22.6 22.5 23.5
                       25.00         1    23.3 22.4 22.1 23.3 21.9 22.7 22.2 22.6
                                     2    22.6 22.6 22.3 23.0 22.1 23.6 22.0 23.0
```

Bobbins are *nested* within plants, but distance interacts with both these factors: the beginning of one bobbin corresponds to the beginning of another. An analysis of variance for this partly nested, partly *crossed* model can be specified in Genstat as follows:

```
TREATMENTS (Plant / Bobbin) * Distance
ANOVA [FPROBABILITY = yes] Force
```

The use of brackets in the treatment model removes an ambiguity about which factors are nested and which interact. For example, the unbracketed model Plant / Bobbin * Distance could alternatively be intended to represent Plant / (Bobbin * Distance) which would mean "the main effects of bobbin and distance and their interaction, within each plant". In fact, Genstat gives higher precedence to "/" than to "*", so the brackets are not strictly necessary here.

4.3 Factors with numerical levels: orthogonal polynomials

The analysis can be taken a stage further. If the breaking force varies significantly between distances, it is likely that this is due to a systematic tendency for strength to increase or decrease along the cord. It would be sensible to perform a regression analysis of strength on distance, and perhaps to include terms for the square and cube of distance, to allow for the possibility that the relationship is curved. Such a regression analysis can be incorporated into the analysis of variance by modifying the TREATMENTS statement as follows:

```
TREATMENTS (Plant / Bobbin) * POL(Distance; 3)
```

The factor Distance is now mentioned as an argument of the function POL, which indicates that this factor is to be broken down into a set of *orthogonal POLynomials*: that is, linear, quadratic, cubic, and perhaps higher-order effects. The second argument, 3, indicates that the analysis is to stop at cubic effects.

The output from such an analysis is potentially voluminous, and the parts that are to be printed can be indicated by setting options in the ANOVA statement, as follows:

```
ANOVA [FPROBABILITY = yes; PRINT = aovtable,contrasts,means; \
    PCONTRASTS = 2; PFACTORIAL = 2] Force
```

The PRINT option indicates the kinds of information to be output, and the FPROBABILITY option indicates that the significance level of each F-statistic in the analysis-of-variance table is to be given. The setting PCONTRASTS = 2 indicates that information is to be printed about the contrasts (that is, the orthogonal polynomials), for the main effects and for two-factor interactions or nested terms, but not for three-factor terms. The setting PFACTORIAL = 2 indicates that the one-way and two-way tables of means are to be printed. The two options used last are both set to 9 by default.

The output from this statement is shown below.

```
14..................................................................

***** Analysis of variance *****

Variate: Force

Source of variation      d.f.       s.s.       m.s.     v.r.  F pr.
Plant                       1    47.4019    47.4019   217.63  <.001
Distance                    5     4.9304     0.9861     4.53  <.001
  Lin                       1     4.5721     4.5721    20.99  <.001
  Quad                      1     0.0964     0.0964     0.44  0.507
  Cub                       1     0.0024     0.0024     0.01  0.917
  Deviations                2     0.2595     0.1298     0.60  0.553
Plant.Bobbin               14    12.3065     0.8790     4.04  <.001
```

```
Plant.Distance              5      2.7869      0.5574     2.56  0.032
  Plant.Lin                 1      0.3064      0.3064     1.41  0.238
  Plant.Quad                1      1.2994      1.2994     5.97  0.016
  Plant.Cub                 1      0.2961      0.2961     1.36  0.247
  Deviations                2      0.8849      0.4424     2.03  0.137
Plant.Bobbin.Distance      70     15.7810      0.2254     1.04  0.434
  Plant.Bobbin.Lin         14      2.6096      0.1864     0.86  0.608
  Plant.Bobbin.Quad        14      5.0312      0.3594     1.65  0.080
  Deviations               42      8.1402      0.1938     0.89  0.658
Residual                   96     20.9100      0.2178
Total                     191    104.1167
```

The Plant and Distance terms in the analysis-of-variance table represent the main effects of these two factors. The Plant.Bobbin term represents the effect of Bobbin-within-Plant, since no main effect of Bobbin has been extracted, whereas the Plant.Distance term represents the Plant-by-Distance interaction, since it is preceded by both the corresponding main effects. The Plant.Bobbin.Distance term represents the Bobbin-within-Plant-by-Distance interaction.

The POL function in the TREATMENTS statement causes each term involving Distance to be broken down into four components. The main effect of Distance comprises a linear component (Lin), representing variation in Force that can be explained by a straight-line regression on distance; quadratic and cubic components (Quad and Cub), representing additional variation that can be explained in terms of distance^2 and distance^3; and a deviation component (Deviations) for variation due to Distance which is not so simply explained. The argument 3 of the POL function indicates that this is not to be resolved into higher-order components. The Lin, Quad, and Cub terms each have one degree of freedom and the remaining degrees of freedom for Distance are assigned to Deviations. Like the degrees of freedom, the sums of squares for Lin, Quad, Cub, and Deviations add up to the total sum of squares for Distance.

The term Plant.Distance is also broken down into components. The component Plant.Lin represents the variation in Force due to change, from one plant to the other, in its linear relationship with distance. Similarly Plant.Quad and Plant.Cub represent variation, between plants, in the quadratic and cubic relationships, and the remainder of the variation in the term Plant.Distance is assigned to Deviations.

Similarly, the linear, quadratic, and cubic relationships can vary from bobbin to bobbin within a plant, and the Plant.Bobbin.Distance term is broken down into Plant.Bobbin.Lin and Plant.Bobbin.Quad terms, which represents this variation. However, the Plant.Bobbin.Cub term is not separated from the Deviations. This is because it is a fifth-order submodel term—that is, the interaction of the third polynomial with two other factors—and submodel terms are treated separately only up to fourth order by default. The Plant.Bobbin.Cub term could, however, be obtained by setting the option CONTRASTS = 5 in the ANOVA statement.

It is sometimes preferable not to estimate higher-order factorial terms but to absorb them into the error, to increase its degrees of freedom. The order beyond which this is done can be specified by setting the option FACTORIAL to the appropriate numbers. Similarly, the option DEVIATIONS indicates the highest-order factorial term in which deviations from submodels are to be distinguished from residual variation.

There are eight bobbins in each plant, so the variation in their linear relationships from the mean relationship for the plant has seven degrees of freedom. This applies to both plants, so that the term Plant.Bobbin.Lin has 14 degrees of freedom altogether. Similar arguments apply to the degrees of freedom of the other components.

The variance ratio (v.r.) for each term is calculated by dividing its mean square by the residual mean square, and is followed by the probability of obtaining so large a value by chance. There are significant effects of Plant, of Distance, attributable to the linear component, of Bobbin-within-Plant, and of Plant-by-Distance, attributable to the quadratic component. The Deviations component of Plant-by-Distance, while not significant, is quite large, suggesting that it might have been worth including a quartic term in the model. On the other hand the Deviations component in the Plant.Bobbin.Distance term has a variance ratio less than 1, suggesting that the Plant.Bobbin.Cub component would not be worth extracting.

The analysis-of-variance table is followed in the output by the table of contrasts shown below.

```
***** Tables of contrasts *****

Variate: Force

*** Distance contrasts ***

Lin      0.0181    s.e. 0.00394    ss.div. 14000.

Quad -0.00036     s.e. 0.000540   ss.div. 746667.

Deviations   e.s.e. 0.0825 ss.div. 32.0

 Distance       0.00      5.00     10.00     15.00     20.00     25.00
               0.005    -0.026     0.055    -0.058     0.031    -0.006

*** Plant.Distance contrasts ***

Plant.Lin   e.s.e. 0.00558 ss.div. 7000.

    Plant          I          II
             -0.0047     0.0047

Deviations   e.s.e. 0.117 ss.div. 16.0

    Plant Distance       0.00      5.00     10.00     15.00     20.00     25.00
        I                0.03     -0.11      0.10      0.02     -0.07      0.03
       II               -0.03      0.11     -0.10     -0.02      0.07     -0.03
```

In order to interpret this table, more needs to be said about the calculation of orthogonal polynomials.

The linear, quadratic, and cubic effects are calculated by multiplying each observation of Force by a coefficient that depends on its level of Distance, and summing the products. The coefficients are as follows:

	Distance (in 100s of yards)					
Effect	0	5	10	15	20	25
Lin	− 12.5	− 7.5	− 2.5	2.5	7.5	12.5
Quad	83.3333	− 16.6667	− 66.6667	− 66.6667	− 16.6667	83.3333
Cub	− 375.0	525.0	300.0	− 300.0	− 525.0	375.0

These coefficients are *polynomials* of Distance because they can be expressed in the following form:

$$k_{\text{lin}} = \text{Distance} - 12.5$$
$$k_{\text{quad}} = \text{Distance}^2 - 25 \times \text{Distance} + 83.3333$$
$$k_{\text{cub}} = \text{Distance}^3 - 37.5 \times \text{Distance}^2 + 342.5 \times \text{Distance} - 375.0$$

where k_{lin}, k_{quad}, k_{cub} are the linear, quadratic, and cubic coefficients respectively. (The rules that determine the numbers used in these polynomials need not concern us here.) They are *contrasts* because each set sums to zero; for example,

$$-12.5 - 7.5 - 2.5 + 2.5 + 7.5 + 12.5 = 0$$

and they are *orthogonal* because the sum of products of any two sets of coefficients is zero; for example:

Lin and Quad

$$\begin{aligned}
&(-12.5 \times 83.3333) + \\
&(-7.5 \times -16.6667) + \\
&(-2.5 \times -66.6667) + \\
&(2.5 \times -66.6667) + \\
&(7.5 \times -16.6667) + \\
&(12.5 \times 83.3333) = 0
\end{aligned}$$

The sum of squares for each effect is given by the formula

$$\text{SS} = \frac{(\Sigma ky)^2}{\Sigma k^2}$$

where:

SS = sum of squares for the effect,
y = the response variable, Force, and
k = the coefficient appropriate to the effect and the distance.

The summation is over all observations in the experiment. SS_{lin} is equal to the sum of squares for a straight-line regression of Force on Distance. SS_{quad} is equal to the increase in the regression sum of squares when Distance^2 is added to the model, and SS_{cub} to the increase when Distance^3 is added. In the table of effects, the contrasts for the main effect of Distance are tabulated first. The estimate of the effect of a contrast is given by the formula

$$\text{Effect} = \frac{\Sigma ky}{\Sigma k^2}$$

the summation again being over all observations in the experiment. The notation "ss.div." stands for sum-of-squares divisor and gives the value of k^2, which is 14000 in the case of the linear effect of Distance. The standard error of a contrast effect estimate is given by the formula

$$\text{SE} = \frac{\text{MS}_{\text{resid}}}{\Sigma k^2}$$

where:

MS_{resid} = residual mean square

In a design with more than one error stratum, such as a split-plot design, the residual mean square from the stratum appropriate to the effect being tested would be used, but in the present case there is only one residual mean square, in the *UNITS* stratum. The significance of the linear effect of Distance can be tested by the formula:

$$t = \frac{\text{Estimate}}{\text{SE}_{\text{Estimate}}} = \frac{0.0181}{0.00394} = 4.59$$

Within the limit of accuracy of the calculation, this value of t is the square root of the corresponding variance ratio in the analysis-of-variance table: 20.99.

Similar information is given for the quadratic component of the main effect of Distance. Printing of the cubic contrast effect is prevented by the option setting PCONTRASTS = 2, and this seems reasonable in view of its low F-value. The value of Force at any given Distance can be estimated by the equation:

$$\text{Force} = \text{grand mean} + k_{\text{lin}} \times \text{Estimate}_{\text{lin}} + k_{\text{quad}} \times \text{Estimate}_{\text{quad}}$$

For example, for the first level of Distance, 0, this formula gives:

$$\begin{aligned} y &= 22.029 - 12.5 \times 0.0181 + 83.3333 \times (-0.00036) \\ &= 22.029 - 0.22625 - 0.03000 = 21.77275 \end{aligned}$$

However, the observed mean for Distance 0 given in the tables of means is 21.775, which deviates from this estimate by 0.00225. The deviations from the full model fitted, including the cubic term, are presented below the contrasts. At Distance 0, the full model actually fits the data less closely than the quadratic model: its deviation is 0.005. Each deviation can be expressed as

$$\text{Dev}_i = \frac{\sum_{j=1}^{32} (y_{ij} - \hat{y}_i)}{\sum_{j=1}^{32} 1^2}$$

where:

Dev_i = the deviation for the ith distance,
j = 1...32 over the different plants, bobbins, and duplicate readings, and
y_i = the estimate of Force for the ith distance.

The sum-of-squares divisor in this case is thus:

$$\sum_{j=1}^{32} 1^2 = 32$$

Precise standard errors cannot be calculated for these deviation effects since they are not independent, but an effective standard error, 0.0825, is given. The deviations are all smaller than the standard error, confirming the conclusion from the analysis-of-variance table that they are non-significant.

The contrasts for the Plant.Distance term follow in the table of contrasts. The effects given show that the Lin estimate for Plant I deviates by −0.0047, and the estimate for Plant II by 0.0047, from the value for the main effect of Distance. The details of how these estimates are derived need not be considered, except to note that only half the observations in the experiment contribute to each, so that the sum-of-squares divisor is 7000, whereas it is 14000 for the linear estimate in the main effect of Distance. Deviations from the Quad and Cub estimate are not printed, since printing was limited to second-order effects by the PCONTRASTS option of the ANOVA statement, and Plant.Quad, the interaction of the quadratic effect of Distance with Plant, is a third-order effect. However, the Plant.Quad variance ratio indicates that this effect is significant and it might be wise to repeat the analysis with the option setting PCONTRASTS = 3.

Thus the estimated value of Force for an observation at Distance 0 from Plant I is:

$$\begin{aligned} y &= 22.029 - 12.5 \times (0.0181 - 0.0047) + 83.3333 \times (-0.00036) \\ &= 22.029 - 0.1675 - 0.0300 = 21.831 \end{aligned}$$

The observed value, given in the tables of means below, is 21.444, which deviates from this estimate by 0.387, and not 0.03 as indicated in the Deviations tabulated for the Plant.Distance contrasts, which take into account the Plant.Quad effect. Thus this effect improves the fit of the model, at least at Distance 0.

```
***** Tables of means *****

Variate: Force

Grand mean   22.029

    Plant          I        II
              21.532    22.526

 Distance       0.00      5.00     10.00     15.00     20.00     25.00
              21.775    21.878    22.066    22.038    22.197    22.222

    Plant   Bobbin             1         2         3         4         5         6
        I                 21.242    21.792    21.608    21.825    21.400    21.125
       II                 22.467    22.717    22.142    22.750    22.358    22.825

    Plant   Bobbin             7         8
        I                 21.575    21.692
       II                 22.150    22.800

    Plant Distance          0.00      5.00     10.00     15.00     20.00     25.00
        I                 21.444    21.337    21.619    21.431    21.525    21.837
       II                 22.106    22.419    22.513    22.644    22.869    22.606


*** Standard errors of differences of means ***

Table                      Plant   Distance          Plant         Plant
                                                    Bobbin      Distance
rep.                          96         32             12            16
s.e.d.                    0.0674     0.1167         0.1905        0.1650
```

The tables of means, and of standard errors of differences of means, complete the output. The tables of means are limited to one-way and two-way tables, as specified in the PFACTORIAL option of the ANOVA statement.

4.4 A factorial design with an additional control

The analysis of some experimental designs is complicated by the fact that one factor becomes irrelevant at a certain level of another factor. For example, in a comparison of the effects of adding either chromium or manganese to steel, it might seem reasonable to consider the metal added, the quantity added, and the metal-by-quantity interaction. But adding no chromium is the same as adding no manganese. This problem occurred in a more subtle way in an experiment described in *Statistics in biology* (Bliss 1970). The yield of beet roots in soil with chips of pine, oak-and-hickory, and aspen-and-birch wood, and without wood chips, was measured at different levels of nitrogen fertilizer. However, the nitrogen was added at rates of 0.0, 0.5, and 1.0 g per 100 g of wood chips, and hence no nitrogen was added in the absence of wood chips. The experiment was planted in a completely randomized design with three replications, except for the no-wood no-nitrogen treatment which was replicated nine times.

The data are presented below.

```
16.9  15.5  16.8  13.3  20.8  21.2  3.6  10.0  18.3  1.2  6.0   7.8
17.5  20.0  19.7  16.1  16.5  18.4  3.8  10.5  18.2  1.2  3.3   9.7
20.0  19.0  20.5  14.3  14.0  16.3  1.8   9.9  14.2  4.1  5.1  10.5:
```

The first three columns represent beet roots grown without wood chips, the next three roots grown with pine chips, the next three with oak-hickory chips, and the last three with aspen-birch chips. Within each group of three columns except the first, the columns represent roots grown with 0.0, 0.5, and 1.0 g nitrogen per 100 g of chips respectively.

The following Genstat program reads the data and creates factors with appropriate values.

```
UNITS [NVALUES = 36]
FACTOR [LABELS = !T(None,Pine,'Oak & Hickory','Aspen & Birch'); \
    VALUES = 3(1...4)3] Species
FACTOR [LEVELS = !(0,0.5,1); VALUES = (3(0),(0,0.5,1)3)3] Nitrogen
OPEN 'Beetyld.dat'; CHANNEL = 2
READ [CHANNEL = 2] Yield
```

Note that in both of the FACTOR statements the directive name FACTOR has been spelled out, rather than using the ampersand (&) the second time. This is because when the ampersand is used to repeat a directive name all the option settings are also repeated unless explicitly changed. Since Nitrogen has no labels, and fewer levels that Species, this would be incorrect. The rather complex number list in the declaration of the factor NITROGEN can be partially expanded to (0,0,0,0,0.5,1,0,0.5,1,0,0.5,1)3: that is, a list of 12 numbers repeated three times.

Although the no-wood no-nitrogen treatment has been assigned to level 0 of Nitrogen, this is not strictly accurate, since it does not represent 0 g/100 g of chips. One way of making the design balanced would be to assign three of the no-wood no-nitrogen observations to 0 g/100 g, three to 0.5 g/100 g, and three to 1 g/100 g. However, this would waste information since the resulting analysis would seek differences that cannot exist. Another way of looking at the categories is shown in Figure 4.1.

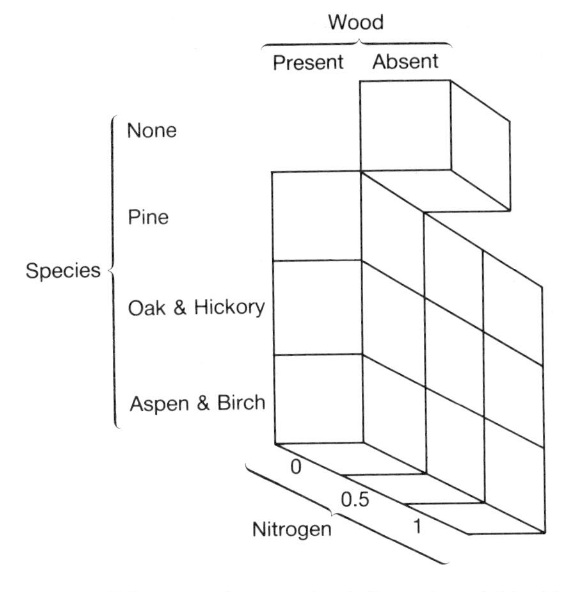

Figure 4.1. Diagram of categories influencing yield of beet.

This diagram illustrates the need for a third factor, Wood, with levels Absent and Present, to describe the design fully. A suitable factor can be declared as follows:

```
FACTOR [LABELS = !T(Absent,Present); VALUES = (3(1),9(2))3] Wood
```

Genstat requires that every observation shall have a value for every factor, but the values of Species and Nitrogen are irrelevant if an observation belongs to the level Absent of Wood. Each such observation is therefore assigned arbitrary values—None and 0—and provided that the same arbitrary values are used for each observation they will not interfere with the analysis.

The experiment can be analysed by the statements

```
TREATMENTS Wood / (Species * Nitrogen)
ANOVA [FPROBABILITY = yes] Yield
```

which produce the following output:

```
10......................................................................

***** Analysis of variance *****

Variate: Yield

Source of variation        d.f.       s.s.       m.s.     v.r.   F pr.
Wood                          1    399.053    399.053   112.56   <.001
Wood.Species                  2    584.827    292.413    82.48   <.001
Wood.Nitrogen                 2    314.229    157.114    44.32   <.001
Wood.Species.Nitrogen         4     77.011     19.253     5.43   0.003
Residual                     26     92.180      3.545
Total                        35   1467.300
```

```
***** Tables of means *****

Variate: Yield

Grand mean   12.67

     Wood     Absent   Present
               18.43     10.74
     rep.          9        27

     Wood  Species            None            Pine Oak & Hickory Aspen & Birch
   Absent                    18.43
  Present                                   16.77          10.03            5.43

     Wood Nitrogen       0.00       0.50      1.00
   Absent               18.43
  Present                6.60      10.68     14.96

     Wood        Species Nitrogen        0.00       0.50       1.00
   Absent           None                18.43
                              rep.          9
  Present           Pine                14.57      17.10      18.63
                              rep.          3          3          3
            Oak & Hickory                3.07      10.13      16.90
                              rep.          3          3          3
            Aspen & Birch                2.17       4.80       9.33
                              rep.          3          3          3


*** Standard errors of differences of means ***

Table                  Wood          Wood          Wood          Wood
                                  Species      Nitrogen       Species
                                                             Nitrogen
rep.                unequal             9             9       unequal
s.e.d.                                                          1.537  min.rep
                      0.725         0.888         0.888         1.255  max-min
                                                               0.888X  max.rep

(No comparisons in categories where s.e.d. marked with an X)
```

The effects of Species, Nitrogen, and their interaction are estimated only within levels of Wood, and since they do not vary within Absent this effectively means within Present. Note that the tables of means give no values for non-existent combinations such as Wood Absent with Species Pine.

4.5 Subdividing the effects of a factor: orthogonal contrasts

In order to look at the effects of Nitrogen and Species in more detail, it is necessary to restrict attention to the treatments where Wood is Present, by means of the statement:

```
RESTRICT Species,Nitrogen,Yield; Wood .EQ. 2
```

The effect of Nitrogen, like that of Distance in the previous example, can then be divided into linear and quadratic components. Higher-order effects cannot be estimated since Nitrogen has only three levels, whose means will be fitted perfectly by a quadratic curve. The new TREATMENTS statement, in which the factor Wood does not appear, is therefore as follows:

```
TREATMENTS Species * POL(Nitrogen; 2)
```

However, in this case we can also be more precise about the nature of the variation between different types of wood chips. The question: "Do types of wood differ?" can be broken down into the questions:

1. Does the softwood pine differ from the hardwoods oak-hickory and birch-aspen?
2. Does one hardwood differ from the other?

Any difference of the first type will be summarized by the expression

$$\text{Effect}_1 = 0 \times y_1 + 2 \times y_2 - 1 \times y_3 - 1 \times y_4$$

and any difference of the second type by the expression

$$\text{Effect}_2 = 0 \times y_1 + 0 \times y_2 + 1 \times y_3 - 1 \times y_4$$

where:

y_1 = mean yield of control,
y_2 = mean yield with pine chips,
y_3 = mean yield with oak-hickory chips, and
y_4 = mean yield with aspen-birch chips.

The coefficients of these expressions, like those of the polynomials in the previous example, sum to zero:

$$0 + 2 - 1 - 1 = 0$$
$$0 + 0 + 1 - 1 = 0$$

Comparisons that can be made by such expressions are called *contrasts*, and these contrasts are *orthogonal* in that the sum of products of their coefficients is zero:

$$(0 \times 0) + (2 \times 0) + (-1 \times 1) + (-1 \times -1) = 0$$

These coefficients can be specified in Genstat in a matrix with two rows and four columns, as follows:

```
MATRIX [ROWS = 2; COLUMNS = 4; VALUES = 0,2,-1,-1, \
                                        0,0, 1,-1] Coef
```

The values of a matrix are always given row by row, and could all have been presented on the same line, but this layout makes them easier for people (as opposed to computers) to read.

The coefficients are introduced into the model as follows:

```
TREATMENTS REG(Species; 2; Coef) * POL(Nitrogen; 2)
```

The factor Species is now an argument of the function REG, which like POL indicates that the factor is to be broken down into components. (The name REG is used because orthogonal contrasts can be regarded as a form of regression analysis.) The second argument, 2, indicates that two orthogonal contrasts are to be evaluated, and the third argument, Coef, indicates the matrix where their coefficients are given.

The PFACTORIAL option in the ANOVA statement is set to 0 to prevent the means being printed again:

```
ANOVA [FPROBABILITY = yes; PFACTORIAL = 0] Yield
```

This statement produces the following output.

```
15..................................................................

***** Analysis of variance *****

Variate: Yield

Source of variation      d.f.       s.s.       m.s.     v.r.   F pr.
Species                    2     584.827    292.413    79.05   <.001
  Reg1                     1     489.607    489.607   132.37   <.001
  Reg2                     1      95.220     95.220    25.74   <.001
Nitrogen                   2     314.229    157.114    42.48   <.001
  Lin                      1     314.169    314.169    84.94   <.001
  Quad                     1       0.060      0.060     0.02   0.900
Species.Nitrogen           4      77.011     19.253     5.21   0.006
  Reg1.Lin                 1      41.388     41.388    11.19   0.004
  Reg2.Lin                 1      33.333     33.333     9.01   0.008
  Reg1.Quad                1       1.080      1.080     0.29   0.596
  Reg2.Quad                1       1.210      1.210     0.33   0.574
Residual                  18      66.580      3.699
Total                     26    1042.647

***** Tables of means *****

Variate: Yield

Grand mean  10.74
```

The mean squares for Species, Nitrogen, and Species.Nitrogen are the same as those for Wood.Species, Wood.Nitrogen, and Wood.Species.Nitrogen in the previous analysis. However, the variance ratios are different, since the values with wood absent no longer contribute to the residual mean square. Each mean square in this analysis could in fact be tested against the residual mean square of the previous analysis, but the conclusions reached would be little altered. The Reg1 component of the Species effect shows that hardwoods differ from the softwood, while the Reg2 component shows that there is a smaller but still significant difference between the hardwoods. The main effect of Nitrogen is almost entirely due to its linear component. In the Species.Nitrogen term, the significant Reg1.Lin component shows that the linear effect of nitrogen varies between hardwoods and the softwood, while the significant Reg2.Lin component shows that it also varies between the two hardwoods. The other two components show that the quadratic effect of nitrogen does not vary significantly between woods.

4.6 Eliminating the effect of a nuisance variable: covariance analysis

Experimenters often take the precaution of measuring additional variables that they believe may influence the variable of most interest. If such *covariates* are influenced by the treatments applied, they may explain the treatment effects observed in the main variable. On the other hand, if they contribute only to error variation, knowledge of them can increase the precision with which treatment effects are estimated. The latter situation occurs in "before and after" trials. In one such trial, described in *Statistical methods* (Snedecor and Cochran 1980), the abundance of leprosy bacilli on patients was estimated on a numerical scale, before treatment with a drug, and again after several months of treatment. Two antibiotics, designated A and D, and one inert control, designated F, were used, each patient receiving one drug. Clearly the drugs could not influence the abundance of bacilli before they were applied, but the initial abundance would be likely to influence the final abundance. The following program analyses these data.

```
UNITS [NVALUES = 30]
FACTOR [LABELS = !T(A,D,F); VALUES = (1...3)10] Drug
OPEN 'Leprosy.dat'; CHANNEL = 2
READ [CHANNEL = 2] Bacilli[1,2]
MATRIX [ROWS = 2; COLUMNS = 3; VALUES =-1,-1,2, \
                                        -1, 1,0] Cont

TREATMENTS REG(Drug; 2; Cont)
COVARIATES Bacilli[1]
ANOVA [FPROBABILITY = yes; UPRINT = aovtable; \
    CPRINT = aovtable] Bacilli[2]
STOP
```

The COVARIATES statement causes the effect of the initial abundance of bacilli, Bacilli[1], to be taken into account in the analysis. The ANOVA statement then produces the following output:

```
10..........................................................................

***** Analysis of variance *****

Variate: Bacilli[1]

Source of variation      d.f.         s.s.         m.s.     v.r.  F pr.
Drug                        2        72.87        36.43     1.66  0.209
  Reg1                      1        70.42        70.42     3.21  0.085
  Reg2                      1         2.45         2.45     0.11  0.741
Residual                   27       593.00        21.96
Total                      29       665.87


10..........................................................................

***** Analysis of variance *****

Variate: Bacilli[2]

Source of variation      d.f.         s.s.         m.s.     v.r.  F pr.
Drug                        2       293.60       146.80     3.98  0.030
  Reg1                      1       290.40       290.40     7.88  0.009
  Reg2                      1         3.20         3.20     0.09  0.771
Residual                   27       995.10        36.86
Total                      29      1288.70


10..........................................................................

***** Analysis of variance (adjusted for covariate) *****

Variate: Bacilli[2]
Covariate: Bacilli[1]

Source of variation      d.f.         s.s.         m.s.     v.r. cov.ef.  F pr.
Drug                        2        68.55        34.28     2.14    0.94  0.138
  Reg1                      1        68.55        68.55     4.27    0.89  0.049
  Reg2                      1         0.06         0.06     0.00    1.00  0.952
Covariate                   1       577.90       577.90    36.01          <.001
Residual                   26       417.20        16.05             2.30
Total                      29      1288.70

***** Covariate regressions *****

Variate: Bacilli[2]

Covariate                coefficient         s.e.
 Bacilli[1]                     0.99        0.164


***** Tables of means (adjusted for covariate) *****

Variate: Bacilli[2]
Covariate: Bacilli[1]

Grand mean  7.90

     Drug         A        D        F
               6.71     6.82    10.16


*** Standard errors of differences of means ***

Table                     Drug
rep.                        10
s.e.d.                   1.846
```

The options UPRINT and CPRINT in the ANOVA statement determine what printing is to be done for the uncorrected analysis of Bacilli[2] and for the analysis of the covariate, Bacilli[1], respectively. In each case, only the analysis-of-variance table is required. The PRINT option controls printing for the analysis of Bacilli[2] corrected for the covariate, but in this case the default setting is satisfactory.

The analysis of variance of Bacilli[1] shows, as expected, that neither the Drug term nor either of its components is significant, though Reg1 comes surprisingly close, presumably by chance. The uncorrected analysis of Bacilli[2] shows that the two antibiotics differ significantly from the control, but not from each other. In order to calculate the adjusted analysis of Bacilli[2] this variable is regressed on Bacilli[1], assuming that thc three levels of Drug have the same slope but different intercepts. Using the regression equation, an estimate is then made of what thc abundance of bacilli would have been on each patient after treatment if they had all had the same abundance to start with. The initial abundance used for this purpose is the mean of Bacilli[1]. The analysis of variance is then recalculated using the adjusted values, a term being added for the regression on the covariate, and the residual sum of squares and degrees of freedom being correspondingly reduced. Since the regression was highly significant the residual mean square is much lower than before. This improvement in the precision of the experiment is indicated by the covariance efficiency factor (cov.ef.) of the residual term, which is thc ratio between the residual mean squares before and after adjusting for the covariate. But this adjustment also inevitably causes some loss of efficiency in the estimation of treatment effects, which is indicated by the covariance efficiency factors of these terms. A value of 1 indicates that a treatment effect is unaffected by the adjustment; a value of 0 indicates that it is completely eliminated by it. In the present case the loss of efficiency is sufficiently great that the variance ratios are reduced by the adjustment, rather than increased as was hoped. Evidently the differences in final abundances of bacilli after treatment with the three drugs were largely due to the chance differences in their initial abundances. An incidental consequence of the adjustment for the covariate is that the two contrasts are no longer precisely orthogonal, and their sums of squares do not add up to the Drug sum of squares.

The coefficient of Bacilli[1] in the regression equation and its standard error are presented under the heading "Covariate Regressions". (In analyses with a BLOCKS statement, separate regressions are performed in each of the strata.) If the coefficient is divided by its standard error, a *t*-statistic of 6.04 is obtained, which is the square root of the variance ratio for the covariate in the analysis of variance.

The table of means and the standard errors at the end of the output are adjusted for covariates: that is, they are estimates of the values that would have been obtained if all patients had started with the same abundance of bacilli.

4.7 Exercises

4(1) Re-analyse the data from Exercise 1(3) in Chapter 1, treating the five explanatory variables as factors, each with two levels, and using TREATMENTS and ANOVA statements. First perform an analysis determining only the main effect of each factor, and confirm that each F-statistic obtained is the square of the *t*-statistic for the corresponding term in the previous analysis. Then attempt to perform an analysis obtaining all the interaction effects. How does the information provided in the output relate to the error message obtained in the previous analysis?

4(2) Measurements were taken of the growth rates (cm/year) of two species of alders and two species of willows at three locations at different distances from the edge of a reservoir. Within each location there were four replications in a completely randomized design. The results are given below. (Unpublished data from C. Gill.)

	Species			
	Alder 1	Alder 2	Willow 1	Willow 2
Location				
1	23.3	7.3	8.0	9.0
	54.4	5.7	6.7	6.3
	25.3	8.0	12.0	2.7
	31.7	8.7	7.6	5.0
2	13.3	7.0	8.0	3.0
	10.4	8.0	8.6	5.1
	17.0	5.1	8.0	7.9
	6.0	5.3	3.4	3.7
3	5.6	2.6	5.5	2.1
	10.0	4.0	8.4	6.0
	17.7	6.7	20.4	5.5
	16.6	2.4	12.3	5.0

Analyse these data using Location and Species as treatment factors, extracting appropriate orthogonal contrasts among the species. Modify the analysis so that the location is treated as a block factor and the effect of species is tested against the location by species interaction.

4(3) In an experiment to test the effect of the proprietary blood substitute Fluosol-DA on immunocompetence in rats, groups of six rats were injected with Fluosol-DA and with sheep's red blood cells, and in each case a control group of three rats was injected with red blood cells from the same batch. The red blood cells were always injected on day 0, but the day of injection with Fluosol-DA varied between groups of rats. Immunocompetence was measured by the number of plaques of lysed cells formed in a standard assay. The results are given below. The same batch number at different days of injection refers to different batches of blood. (Unpublished data from A. Bollands.)

Day of injection with Fluosol-DA	Batch	Rat Number Fluosol-DA treated group 1	2	3	4	5	6	Control group 1	2	3
−4	1	95	190	890	122	350	332	81	515	918
	2	74	239	439	186	103	183	429	232	177
	3	64	316	56	117	58	551	602	169	305
−1	1	22	342	32	4	*	*	247	4652	1108
	2	366	131	124	42	81	*	100	696	241
	3	282	546	539	250	1173	755	380	658	174
0	1	670	214	824	205	158	2132	169	48	253
	2	325	459	411	1866	413	983	109	62	183
	3	1064	567	722	452	5172	4007	879	962	722
+1	1	1084	158	115	215	440	1119	110	1635	684
	2	523	320	1073	1140	*	*	801	967	700
	3	329	759	3140	1042	2254	218	988	661	1039

Obtain the square roots of these values. Analyse both the untransformed and transformed values so as to obtain the effects of day of injection and of batch within day of injection, and the interaction of these with the comparison between control and treatment. Divide the day-of-injection term into orthogonal polynomial terms. Test graphically whether the square root transformation decreases the non-Normality and heterogeneity of the residuals, as would be expected for data following a Poisson distribution.

For both the untransformed and transformed values, obtain a table of the means for each treated and untreated group. Use an EQUATE statement to place these means in a variate, then use a CALCULATE statement to obtain a variate containing the differences between the means for the treated groups and the means for the corresponding untreated groups. (Qualified identifiers can be used in a CALCULATE statement to restrict the calculation to certain values. Thus the statement

```
CALCULATE A = B$[!(5...8)]
```

causes only the fifth, sixth, seventh, and eighth values of B to be placed in A.) Perform an analysis of variance to test the effect of day of injection on these differences.

5 The analysis of variation in several variables

5.1 Introduction

In this chapter we describe two commonly used multivariate methods, principal components analysis and canonical variates analysis, and the way in which Genstat can be used to apply them. The former is used to examine the interrelationships of a set of variates, and the units for which the variates have been measured. The latter is a form of discriminant analysis and is used to assess differences among groups of units. Both of these methods can be derived from a classical approach that operates in terms of sums-of-squares-and-products (SSP) matrices among the variables involved. Both methods can also be viewed in terms of points representing the units, and the groups in the case of canonical variate analysis. This is a geometric approach which ties in very neatly with the methods described in the following two chapters. Before describing the methods in detail, three general comments about multivariate analysis are necessary.

Many univariate methods, for example the analysis of variance, are based on particular statistical distributions, commonly the Normal distribution. In multivariate analysis this is not generally the case: indeed all the methods described in this chapter and the next two need no underlying distributional assumptions. However, in some instances it is useful to make some assumptions about the distribution from which the data come, provided that these assumptions are valid!

In order to interpret any multivariate analysis, you must at some stage be prepared to consider the data in terms of a set of points in several dimensions. With luck the number of dimensions can be reduced to (at most) two, so that the points can be plotted on a piece of paper. Unfortunately this is not always the case; for example, in the principal components analysis given below three dimensions are used.

Preliminary transformation of the data can often make an important difference to the results of multivariate analysis. A transformation that is of particular value is the use of logarithms: we shall mention this again in Section 5.4. Unlike their use in univariate statistics, transformations of multivariate data are not usually intended to improve the validity of distributional assumptions, such as Normality.

The first and last few lines of a file of data on life expectancy in 25 countries, taken from *Clustering Algorithms* (Hartigan 1975) and originally published in *Population: facts and methods for demography* (Keyfitz and Flieger 1971), are set out below. On each line the name of a country is given followed by a two-character code. The next four variables give the life expectancy for males at ages 0, 25, 50, and 75; the last four variables are for females at the same ages.

Algeria Al	63	51	30	13	67	54	34	15
Cameroon Cm	34	29	13	5	38	32	17	6
Madagascar Md	38	30	17	7	38	34	20	7
. . .								
Ecuador Ec	57	46	25	9	60	49	28	11

Reading the data from a secondary file gives the Genstat output shown below. The list of identifiers of the eight variables will be needed in several places so a pointer is declared on Lines 3 and 4 to represent them.

```
 1  UNITS [NVALUES=25]
 2  TEXT Country,Code
 3  POINTER [VALUES=Male0,Male25,Male50,Male75, \
 4                  Female0,Female25,Female50,Female75] Vars
 5  READ [CHANNEL=2; PRINT=data,summary] Country,Code,Vars[]

 1  Algeria  Al                   63 51 30 13   67 54 34 15
 2  Cameroon  Cm                  34 29 13  5   38 32 17  6
 3  Madagascar  Md                38 30 17  7   38 34 20  7
 4  Mauritius  Mr                 59 42 20  6   64 46 25  8
 5  Reunion  Ru                   56 38 18  7   62 46 25 10
 6  Seychelles  Sy                62 44 24  7   69 50 28 14
 7  Tunisia  Tu                   56 46 24 11   63 54 33 19
 8  Canada  Cn                    69 47 24  8   75 53 29 10
 9  'Costa Rica'  CR              65 48 26  9   68 50 27 10
10  'Dominican Rep.'  DR          64 50 28 11   66 51 29 11
11  'El Salvador'  ES             56 44 25 10   61 48 27 12
12  Greenland  Gl                 60 44 22  6   65 45 25  9
13  Grenada  Gd                   61 45 22  8   65 49 27 10
14  Guatemala  Gt                 49 40 22  9   51 41 23  8
15  Honduras  Hn                  59 42 22  6   61 43 22  7
16  Jamaica  Jm                   63 44 23  8   67 48 26  9
17  Mexico  Mx                    59 44 24  8   63 46 25  8
18  Nicaragua  Nc                 65 48 28 14   68 51 29 13
19  Panama  Pn                    65 48 26  9   67 49 27 10
20  Trinidad  Ti                  64 43 21  6   68 47 24  8
21  USA  US                       67 45 23  8   74 51 28 10
22  Argentina  Ag                 65 46 24  9   71 51 28 10
23  Chile  Ch                     59 43 23 10   66 49 27 12
24  Columbia  Co                  58 44 24  9   62 47 25 10
25  Ecuador  Ec                   57 46 25  9   60 49 28 11
26  :

  Identifier   Minimum      Mean   Maximum    Values   Missing
       Male0     34.00     58.92     69.00        25         0
      Male25     29.00     43.64     51.00        25         0
      Male50     13.00     23.12     30.00        25         0
      Male75     5.000     8.520    14.000        25         0
     Female0     38.00     63.16     75.00        25         0
    Female25     32.00     47.36     54.00        25         0
    Female50     17.00     26.32     34.00        25         0
    Female75      6.00     10.28     19.00        25         0
```

5.2 Principal components analysis

Principal components analysis (PCP) is often used simply to examine the interrelationships among several variables and can be done from the sums-of-squares-and-products (SSP) matrix for the variables. To declare, form, and print the SSP matrix the following Genstat statements are used.

```
   6  SSPM [TERMS=Vars[]] Varssp
   7  FSSPM [PRINT=sspm] Varssp

*** Degrees of freedom ***

Sums of squares:   24
Sums of products: 23


*** Sums of squares and products ***

Male0         1        1601.8
Male25        2         893.3         635.8
Male50        3         512.2         412.1         308.6
Male75        4         146.0         168.7         148.4         114.2
Female0       5        1722.3         935.4         511.5         142.9
Female25      6         916.7         607.2         373.9         168.3
Female50      7         489.6         378.9         254.0         137.8
Female75      8         188.6         195.5         145.2         103.4

                            1             2             3             4

Female0       5        1927.4
Female25      6        1029.6         683.8
Female50      7         554.7         432.1         309.4
Female75      8         246.9         259.5         207.8         191.0

                            5             6             7             8

*** Means ***

Male0               58.92
Male25              43.64
Male50              23.12
Male75              8.520
Female0             63.16
Female25            47.36
Female50            26.32
Female75            10.28

*** Number of units used ***

          25
```

A brief examination of the values of Varssp shows two things. First, the variance of the variables decreases as age increases; second, all the sums of products are positive, and so all the variables are positively correlated. Neither of these features is very surprising: over a range of countries one would expect that life expectancy at birth is considerably more variable than life expectancy at, say, age 50. Similarly, all eight variables are likely to be lower than average in a country with low overall life expectancy, and vice versa.

The total variation in the data is the sum of the diagonal values of Varssp; this is usually termed the *trace* (from the trace operator in matrix algebra which sums the values on the leading diagonal of a square matrix). Principal components analysis partitions the total variation into components for each of a set of new variables. It does this in such a way that the first new variable explains as much of the trace as possible; that is, the values for the different countries have the maximum possible variance. The second new variable must be independent of (which means that it has zero covariance with) the first; subject to that constraint, it must explain as much as possible of what is left of the trace after the first variable has been obtained. Subsequent new variables are similarly derived, and each must be independent of those preceding it. The amount of variation explained by each new variable is given by the value of a latent root (also called an *eigenvalue*). The new variables are formed as linear combinations of the old variables: the coefficients used indicate the weight given to each of the old variables in forming the new one, and are therefore often called *loadings*.

The Genstat directive for principal components analysis is PCP. This directive has three options, of which the first is PRINT; the setting 'roots' corresponds to the latent roots and the trace, and 'loadings' is used to obtain the coefficients of the original variables that are used to form the new variables. After the options the identifier of the SSP matrix is all that is required.

```
   8  PCP [PRINT=roots,loadings] Varssp

 8...............................................................................

 *****  Principal components analysis  *****

 ***  Latent Roots  ***

                    1              2              3              4              5
               5030.3          513.0          163.6           28.4           15.8
                    6              7              8
                  9.5            7.3            4.2

 ***  Percentage variation  ***

                    1              2              3              4              5
                87.15           8.89           2.83           0.49           0.27
                    6              7              8
                 0.16           0.13           0.07

 ***  Trace  ***

       5772

 ***  Latent Vectors (Loadings)  ***

                      1            2            3            4            5
      Male0    -0.55161      0.32868      0.25629     -0.14033     -0.06043
     Male25    -0.33277     -0.26551      0.43948      0.55121      0.00849
     Male50    -0.19852     -0.35152      0.50297     -0.15684      0.39083
     Male75    -0.07126     -0.35786      0.15202     -0.76649     -0.27761
    Female0    -0.60387      0.37472     -0.32122     -0.15602      0.20814
   Female25    -0.35204     -0.27823     -0.30452      0.14040     -0.39542
   Female50    -0.20430     -0.40069     -0.25125      0.13789     -0.38647
   Female75    -0.10087     -0.43757     -0.45392     -0.03616      0.64609
```

```
                 6          7          8
   Male0   -0.61483    0.34357   -0.05105
  Male25    0.38541    0.25995   -0.32498
  Male50   -0.10849   -0.48833    0.39556
  Male75    0.28125    0.22256   -0.22482
 Female0    0.44740   -0.33218   -0.11651
Female25    0.10398    0.20463    0.69077
Female50   -0.37318   -0.49935   -0.42142
Female75   -0.17981    0.35330   -0.12816
```

The latent roots are also expressed as percentages of their total, the trace (5772). Most (87%) of the variation in the data is explained by the first variable; the matrix of latent vectors shows what this variable represents.

Each new variable, whose loadings are given in the corresponding column of the latent vectors matrix, is a linear combination of the eight original variables. From the output you can see that the variable corresponding to the first latent root is

$$\begin{aligned} z_1 = & -0.55 \times x_1 - 0.33 \times x_2 - 0.20 \times x_3 \\ & -0.07 \times x_4 - 0.60 \times x_5 - 0.35 \times x_6 \\ & -0.20 \times x_7 - 0.10 \times x_8 \end{aligned}$$

Here $x_1, x_2 \ldots x_8$ correspond to the eight original variables. However, their values are expressed as deviations about their means, because the sums of squares and products are calculated about the means of the input variables. Thus x_1 contains the life expectancies for males at birth minus the mean of those values, which is roughly 59. Looking at the first column of the latent vectors you can see that all the signs of the coefficients are the same (negative). Therefore z_1 can be interpreted as giving a (negative) overall measure of life expectancy in each country. Note that more emphasis is put on the data from younger ages; however, the sexes are weighted fairly equally.

The reason why the loadings for the first latent vector decrease as age increases is very likely to be because the younger age variables have greater variance, as noted earlier. Since z_1 explains as much variation as possible, it usually has higher loadings for the variables with larger variance. When all the variables are positively correlated, as in this case, all the loadings for the first latent vector will always have the same sign.

The second and third latent vectors are also fairly easy to interpret. The values in the third vector for the male variables are all positive, and those for the female variables are all negative. It is well known that females tend to outlive males by about three years (for these data the average difference is 3.24 years). If a country gets a relatively high value for this new variable it means that the differential life expectancy is less than 3.24; analogously, for relatively low scores the differential is larger than 3.24. The pattern of positive and negative signs also indicates how the second vector can be interpreted. A low value for a country will occur if the life expectancy at birth is relatively low, compared to the other variables, so the second vector is some (inverse) expression of excessive mortality at a young age.

5.3 Principal component scores

Although the output above shows how to form each of the new variables, for example z_1, it has not given the values of these variables for each of the 25 countries. These are called the *principal component scores*, or coordinates. To get the PCP directive to print them the setting 'scores' must be used. A glance at the earlier PCP output shows that the first three latent vectors explain nearly 99% of the variation in the data. Therefore, in terms of showing this variability, only the first three score variables need be printed. To restrict the output to that corresponding to the first three latent roots the second option of PCP is used. This is the NROOTS option. In order to plot the scores, as well as having them printed, they need to be saved from the analysis. Since only the first three score variables are needed they can be held in a matrix with three columns: one for each variable. The SCORES parameter of the PCP statement is then used to save the scores.

```
 9  MATRIX [ROWS=Country; COLUMNS=3] PCscores
10  PCP [PRINT=scores; NROOTS=3] Varssp; SCORES=PCscores

10..........................................................................

*****  Principal components analysis  *****

***  Principal Component Scores  ***
```

	1	2	3
1	-13.086	-10.186	1.094
2	43.814	0.966	-1.402
3	38.921	-2.302	0.562
4	1.772	4.680	-1.142
5	6.089	3.477	-4.788
6	-7.060	0.304	-3.631
7	-4.011	-11.183	-6.499
8	-16.471	4.214	-1.643
9	-9.373	0.586	2.593
10	-9.681	-3.959	3.908
11	1.779	-4.258	0.128
12	-0.195	3.990	0.529
13	-3.140	0.986	-0.646
14	17.362	-2.531	3.079
15	4.956	5.327	2.949
16	-4.659	3.423	0.297
17	0.773	1.652	2.374
18	-11.190	-4.299	2.191
19	-8.417	0.489	3.219
20	-4.081	7.328	-0.256
21	-12.991	5.021	-2.356
22	-10.679	2.265	-0.811
23	-2.518	-0.708	-2.459
24	1.304	-0.562	1.379
25	0.781	-4.719	1.330

There are two relatively large scores in the first column, 43.8 and 38.9, which are for Cameroon and Madagascar, respectively. This column corresponds to the (inverse) overall life expectancy measure and these large values indicate low overall life expectancy. The loadings for the first vector emphasized the life expectancies at birth, and the values of these for both Cameroon and Madagascar are all very low—in fact not much better than at age 25. The best scores (best, that is, in terms of life expectancy) are those for Canada, Algeria, and the USA. The score for Algeria may be surprising; however, comparison of the values for Algeria with the output from the READ statement shows that it has the highest life expectancy values for both Males and Females at ages 25 and 50.

It is usually easier to inspect the scores by plotting them, especially when many units are analysed. This is often done by plotting the score variables in pairs against each other. However, here there is one score variable that contains a great majority of the variation, and two less important variables. In these situations it is often just as good to plot the first variable on its own, by plotting a constant against it, and then to plot the second and third together.

Since the GRAPH directive cannot plot individual columns of matrices, it is necessary to copy the scores out of the matrix PCscores into variates. These are first declared and a CALCULATE statement then uses a qualified identifier to extract the values for all rows, indicated by the * within the square brackets, and for columns 1, 2, and 3 for the three variates, respectively.

The y-variate in the GRAPH statement is the unnamed variate of 25 zero values; in plotting this only a few rows are required for the graph, and the number has been set to 3. The SYMBOLS parameter is used to specify that the two-letter codes are to be used to represent the points.

```
11  VARIATE [NVALUES=Country] Score[1...3]
12  CALCULATE Score[] = PCscores$[*; 1...3]
13  GRAPH [TITLE='First principal component score'; NROWS=3] \
14      !((0)25); Score[1]; SYMBOLS=Code
```

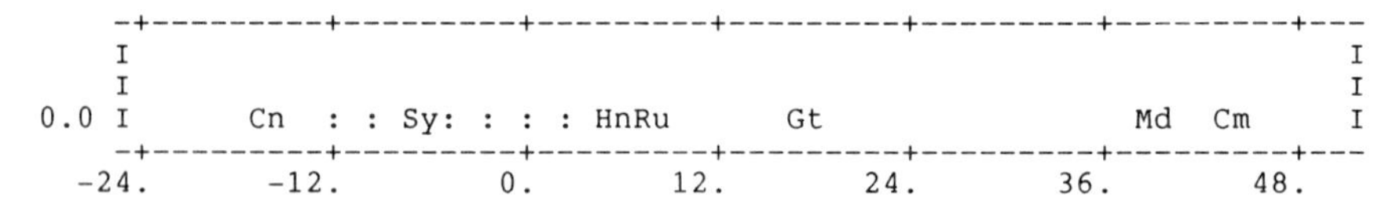

```
                        First principal component score

      -+---------+---------+---------+---------+---------+---------+---
       I                                                               I
       I                                                               I
   0.0 I      Cn   : : Sy: : : : HnRu      Gt                Md  Cm    I
      -+---------+---------+---------+---------+---------+---------+---
    -24.      -12.        0.       12.       24.       36.       48.
```

The high scores for Madagascar and the Cameroon are now made very evident, and the point for Canada is the left-most on the plot. Unfortunately, only seven of the 25 points are clearly labelled; the other 18 points each occur at one of the six colons on the graph, which denote coincident points.

There is no easy way to display the points graphically so that all 25 of them are clearly labelled; however, the values for the countries can be shown in order by sorting them, and the country names, into ascending order. The SORT statement on Line 15 does this, using the values from the variate Score[1] to define the order, and placing the resulting values and country names in the structures Sortscor and Sortname. In the PRINT statement that follows, the ORIENTATION option has been set to specify that the values of the structures, printed in parallel, are to go across the page, rather than down it. The identifiers of the structures are not needed, so the IPRINT option has been set to suppress them. The PRINT directive will determine a suitable number of columns in which to print the values; however, it is useful here to specify that a single decimal place is required—this will be ignored, of course, for the names of the countries.

```
15  SORT Score[1],Country; Sortscor,Sortname
16  PRINT [ORIENTATION=across; IPRINT=*] Sortname,Sortscor; DECIMALS=1
           Canada          Algeria          USA        Nicaragua        Argentina
            -16.5            -13.1        -13.0            -11.2            -10.7
   Dominican Rep.       Costa Rica       Panama       Seychelles          Jamaica
             -9.7             -9.4         -8.4             -7.1             -4.7
         Trinidad          Tunisia      Grenada            Chile        Greenland
             -4.1             -4.0         -3.1             -2.5             -0.2
           Mexico          Ecuador     Columbia        Mauritius      El Salvador
              0.8              0.8          1.3              1.8              1.8
         Honduras          Reunion    Guatemala       Madagascar         Cameroon
              5.0              6.1         17.4             38.9             43.8
```

To plot the third principal component score against the second is fairly straightforward. However, in plots such as this it is important that the scaling, in units per inch, must be the same for each axis. This means that the same mathematical distance is represented by the same physical distance on the plot in any direction. The importance of this will become apparent when we take an alternative view of PCP. The same scaling on each axis is quite simply obtained with the option setting EQUAL = scale in the GRAPH statement. When using this option setting it does not matter what physical size the frame is on the paper; however, it is often convenient to use a physically square graph, in which case suitable values for the NROWS and NCOLUMNS options must be set. Most line-printers print 10 characters per inch across the paper and six lines per inch down the paper; in which case, setting NROWS to $(6t + 1)$ and NCOLUMNS to $(10t + 1)$ will result in a graph t inches square.

```
17  GRAPH [YTITLE='Third principal component score'; \
18     XTITLE='Second principal component score'; EQUAL=scale; \
19     NROWS=37; NCOLUMNS=61] Score[3]; Score[2]; SYMBOLS=Code
```

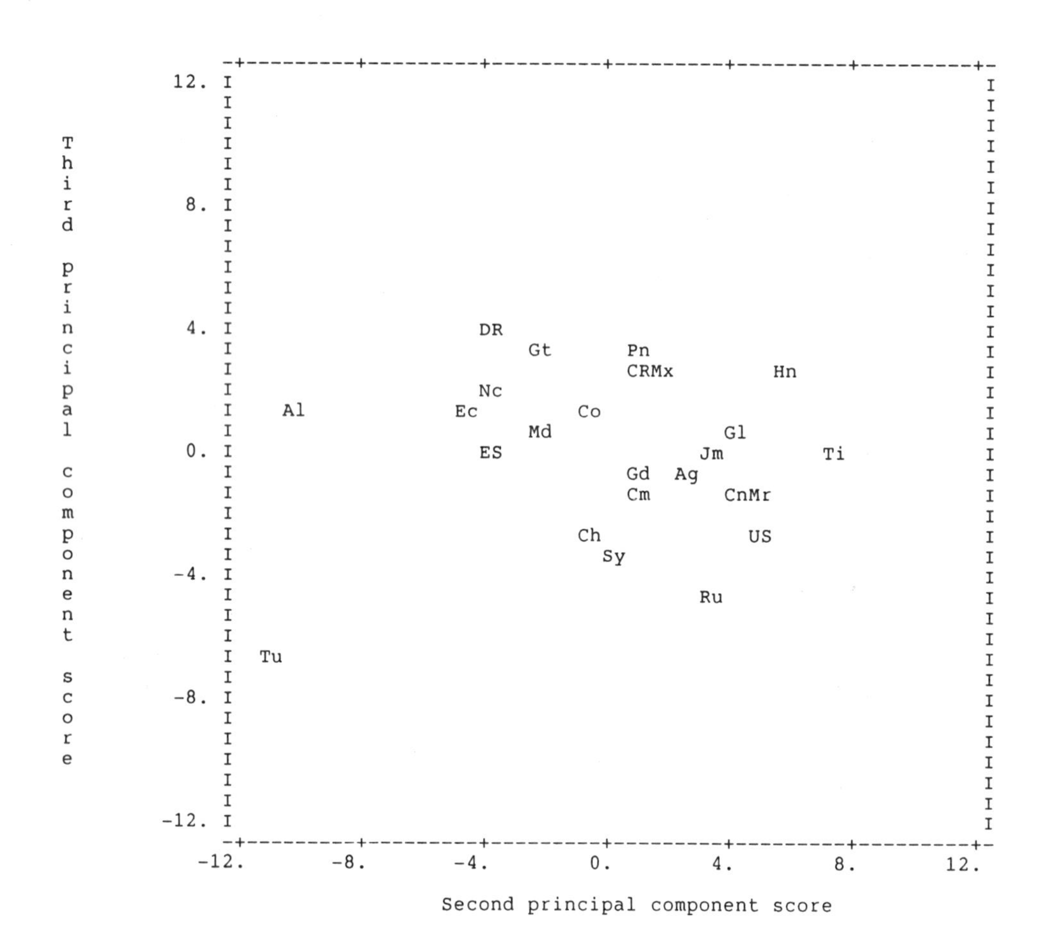

5.4 Principal components analysis: an alternative view

Now we shall move on to a different way of looking at principal components analysis. This gives the same results but, by taking an alternative view of the data, allows a more general approach to multivariate analysis. Rather than considering our data as eight separate variates, think of the data as eight values measured for each of 25 countries. If each value is considered as giving one coordinate of a point for a country then the data can be thought of as giving the coordinates of 25 points, one for each country, in a multi-dimensional space, here of eight dimensions (one for each measurement).

It is hard to think about things in eight dimensions; in fact three dimensions is often difficult enough! In order to simplify the eight-dimensional situation one could ask the question: "do the points in eight dimensions actually lie in, or very close to, a space of fewer dimensions?". To provide an example, consider a set of points in three dimensions, for example in a room. Now if the points all lie close to a plane within the room, for example an artist's easel, the points could be projected onto the easel. This would give a more convenient graphical representation, because a two-dimensional graph could be drawn showing where the projections of the points lie. With even more good fortune the points would lie very close to some straight line within the room, so that the three-dimensional representation could be reduced to a one-dimensional one. Of course the straight line would probably not be parallel to the floor of the room, nor to any of the sides so that the one-dimensional representation would need contributions of information from all three dimensions. Similarly the projection onto an artist's easel, used for a two-dimensional representation, would need pairs of contributions from all three dimensions.

To reduce the dimensionality in this way obviously requires finding the "best" plane, or line, to project the points onto. "Best" often means the one that minimizes the sum of squared residuals, as it does here. Projecting each point onto a plane implies that there is a distance, or *residual*, between the point and its projected position; this distance is naturally measured at right-angles, or orthogonal, to the plane. The residual is then a useful indication of how inaccurate the two-dimensional represention is for that point.

This reduction of dimensionality is precisely what principal components analysis does. By taking data from several variates, in this example eight, and reducing them to a set of three scores we have used PCP to reduce our eight-dimensional representation to a three-dimensional one. The reason why this approach is equivalent to the earlier one, of maximizing variation explained, is quite simple. Since there is a total amount of variation in the data, which is being split up into that which is explained and that which is not, or in other words is residual, maximizing the former must be the same as minimizing the latter.

Earlier we went to the trouble of declaring and forming the SSP structure Varssp which was passed to the PCP directive. In fact this is not necessary: it is adequate to use the variates directly in the PCP directive. However, they must be provided as a single structure, because all the data are being considered as a single dataset. The single structure concerned is a pointer, and its values must be the variates for the analysis.

Before looking at another analysis of the life expectancy data it may be worthwhile to take a logarithmic transformation of the data. This should help to make the variances of the variables more similar, and we would then expect the loadings for the first score to be more nearly equal. When dealing with strictly non-negative data, as here, it is common for larger numbers to be inherently more variable than smaller numbers and taking logs generally gets round this. With univariate data it is common to take logs of counts: this is often said to be done to equalize the variance of the data values and may also be associated with an assumption of underlying Normality on the transformed scale. Here the transformation is being used to make variability more equal *across* a set of variates, rather than within each one, and assumptions about (multivariate) Normality do not enter into the issue at all. A final word about log transformations: most people are more at home with logs to the base 10, rather than the mathematicians' preferred logs to the base *e*, so it is usually better to use the former. The relevant function in Genstat is LOG10.

Only a few more statements are needed to analyse the log-transformed data. It is sometimes useful to look at the loadings in a graphical form, rather than as printed by the PCP directive. As with the scores for the countries, this involves saving the matrix of loadings. The loadings are part of a compound structure which is also used to hold the roots and trace, so it is necessary to declare this LRV structure before doing the analysis. The LRV has two size attributes: the first is the number of rows of the matrix of loadings; the second is its number of columns, which is also used to specify the size of the diagonal matrix of latent roots. Since the rows of the matrix of loadings correspond to the variates represented by the pointer Vars, this can be used to specify the first size attribute of the LRV structure. Having seen the earlier analysis, we can make an educated guess that three dimensions will be sufficient.

```
  20  CALCULATE Vars[] = LOG10(Vars[])
  21  LRV [ROWS=Vars; COLUMNS=3] PClrv
  22  PCP [PRINT=roots] Vars; LRV=PClrv; SCORES=PCscores

22...................................................................

*****  Principal components analysis  *****

***  Latent Roots  ***

                    1           2           3           4           5
               0.9079      0.1948      0.0855      0.0144      0.0051
                    6           7           8
               0.0021      0.0006      0.0004

***  Percentage variation  ***

                    1           2           3           4           5
                74.98       16.09        7.06        1.19        0.42
                    6           7           8
                 0.18        0.05        0.04

***  Trace  ***

    1.211
```

However, just to make sure, the PCP statement will print the latent roots for all eight dimensions, though the values for only the first three will be saved.

The output from PCP shows that the first three roots explain 98% of the total sum-of-squares, giving a residual sum-of-squares of only 2%. Thus, as expected, the first three dimensions are sufficient. To examine the corresponding latent vectors the relevant part of the LRV structure is printed. This is the first element of the LRV, but it also has the label 'Vectors', as used below.

```
  23  PRINT PClrv['Vectors']

              PClrv['Vectors']
                      1          2          3
        Vars
       Male0    -0.2834     0.4866     0.0881
      Male25    -0.2681     0.2243     0.1431
      Male50    -0.3535     0.0935     0.3684
      Male75    -0.4679    -0.5491     0.5448
     Female0    -0.2893     0.4950    -0.0750
    Female25    -0.2630     0.1832    -0.0937
    Female50    -0.3125     0.0350    -0.1430
    Female75    -0.5033    -0.3506    -0.7102
```

The first latent vector is, as before, a *size* vector, with all coefficients having the same sign. The second latent vector looks somewhat different from that in the previous analysis. It distinguishes countries that have a relatively high life expectancy at birth but relatively low expectancy in old age: it will probably bring to the fore those countries in which only a very small proportion of the population reaches 75, since those that do reach that age will have some remaining life expectancy, despite the low expectancy at earlier ages. The third latent vector again shows countries where the relative life expectancy of males and females differs from the norm. It is likely that the second and third dimensions will contain roughly the same information as the second and third dimensions of the previous analysis, but the split of information is slightly different.

Comparing the first latent vector with that from the previous analysis shows that the emphasis has been removed from the data at birth and, if anything, has shifted to the data for older ages. This suggests that the log transformation has over-compensated for the unequal variances: had it been important to even out the variances, taking square roots would probably have been about right, because this is a half-way house between no transformation and taking logs. Alternatively, we could have used the correlations, rather than the sums of squares and products, among the variates. This is done by setting the METHOD option of PCP to 'correlations', as opposed to its default setting of 'ssp'. The PCP directive will then calculate the latent roots and vectors of the matrix of correlations and use these to calculate the scores.

Some people prefer to use correlations most of the time. They certainly have the advantage that there is no uneven weighting of the variables. However, with data such as these, which are all measured in the same units, it seems preferable not to do so. An added disadvantage of the use of correlations is that it is data-dependant, so that the standardization changes if the data for one country are left out. Our own preference is to use correlations only when there is a very good case for doing so, rather than the reverse.

5.5 Biplots

Biplots were introduced by Gabriel (1971) as a useful device to aid the interpretation of principal component analyses. In fact there are several forms of biplot which are described with examples by Gabriel (1981); however, we shall look at the most commonly used form. The underlying idea of the biplot is to show, in a single graphical frame, points for both the units, and the variates, used to define the principal components. The coordinates of the points for the units are the scores from the analysis. The coordinates of the points for the variates are the values of the latent vectors, or loadings. For example, the three-dimensional coordinates of the point for the males-at-birth variate from the last analysis are (− 0.29, 0.49, 0.08). Before we describe how a biplot can be useful, we shall set about constructing one from the saved results of the last analysis.

It is clear that the first set of loadings, although explaining much of the variation in the data, simply represents some average measure of life expectancy. It is likely that a more useful biplot will be for the second and third dimensions (although there is no reason, other than the consideration of space in this book, why one should not produce several biplots, for example for the first and second dimensions).

Before using the GRAPH directive to produce the biplot it is a good idea to check if the scores for the countries and the loadings for the variates are on commensurable scales. The values of the loadings will always be in the range [− 1, + 1]. Although the scores for the countries from the last analysis are not printed above it is possible to deduce roughly what range they lie in. This is because each latent root is the sum of squares of the relevant values of the scores, and each column of scores always has zero mean. For example, the sum of squares of the second set of scores is 0.195. Since there are 25 values, with sum of squares 0.195 it seems likely that none of them is outside the range [− 0.25, + 0.25]. When the scores and loadings are not within similar ranges the usual device is to multiply either the scores or the loadings, normally the latter, by some appropriate constant before producing the biplot. For the graph shown below the coordinates for the variates were divided by three.

```
24  CALCULATE Score[2,3] = PCscores$[*; 2,3]
25  VARIATE [NVALUES=8] Vector[2,3]
26  CALCULATE Vector[] = PClrv[1]$[*; 2,3]/3
27  GRAPH [TITLE='Biplot (dimensions 2 and 3)'; \
28     YTITLE='Third principal component score'; \
29     XTITLE='Second principal component score'; EQUAL=scale; \
30     NROWS=37; NCOLUMNS=61] Score[3],Vector[3]; Score[2],Vector[2]; \
31     SYMBOLS=Code,!T(M0,M1,M2,M3,F0,F1,F2,F3)
```

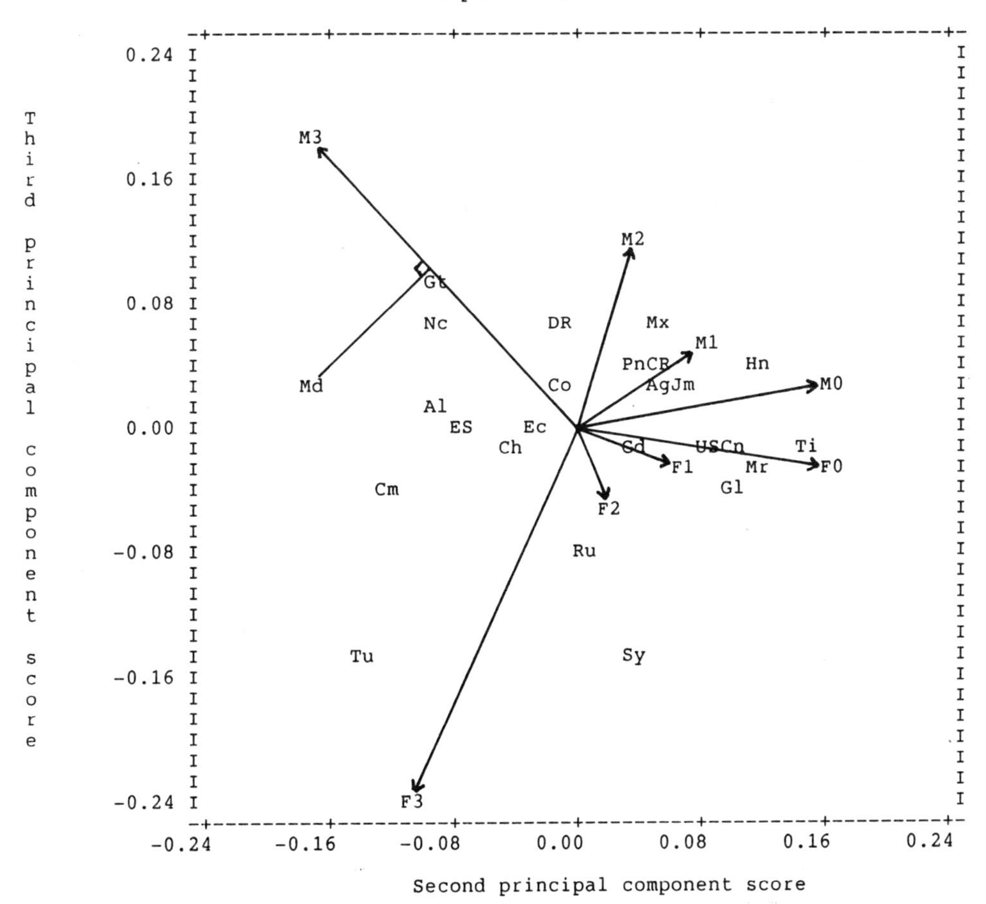

It is usual with biplots to represent the variates as vectors from the origin to the plotted points. This is easiest to do by hand, which is what we have done here. It is clear from the graph that the loadings in the third dimension for the male variates are all positive, and those for the female variates are all negative, highlighting the major aspect of this dimension. Also the way that the two sets of points, for male and female variates, fan around from the right to the left shows that the second dimension is a contrast of the two older variates with the other six.

The biplot can be used to recover aspects of the individual data values. The approximate order of the data for any particular variate can be recovered from the biplot. This is done by projecting the points for the countries onto the vector for the variate concerned. For example, the projection of the point for Madagascar onto the vector for the last male variate is shown on the graph. The approximate order of the data for the 25 countries is the order of the projected points; thus Guatemala (point Gt) should have the highest score on this variate, closely followed by Madagascar and Nicaragua (point Nc). The lowest score is probably for Seychelles (point Sy); for this country, and others, the projection must be made onto a continuation beyond the origin of the vector from the point M3. Although this has not been shown on the graph, such projections are fairly easy to make by eye.

For this biplot, of the second and third dimensions, it must be borne in mind that the overall measure of life expectancy, contained in the first dimension, is not shown. Therefore, any interpretation drawn from the graph must be tempered with what is known from the first dimension. For example, Madagascar does not have a very high life expectancy for males at age 75; however, the value is high, if one takes into account the very low overall life expectancy in this country. The expected ages at death for Madagascan males at different ages is shown below:

age	0	25	50	75
expected age at death	38	55	67	82

The value 82 above is surprisingly high, considering the other values in the table: it is this aspect that the biplot brings out. In a similar way, it can be deduced that countries with points to the right of the graph do not have much trouble, relatively, with infant mortality, since their values will be quite high on the males-at-birth and females-at-birth variates.

5.6 Canonical variate analysis

We have shown how PCP allows the display of the important features of data from a large number of dimensions in a smaller dimensionality. Suppose, however, that the units fall into groups. Now there may be more interest in showing differences between the groups than between the individuals. An example by Delaney and Healy (1966) concerns measurements on white-toothed shrews which come from 10 island populations. The interest is in differentiation of the populations, rather than the individual shrews. The countries from the PCP example can be grouped naturally into their respective geographical regions: Africa, the Indian ocean, North America, Central America, the Caribbean, and South America. The first two of these can be loosely thought of as African and the last four as forming the Americas.

One way of analysing the groups, rather than the units, is to calculate group means for all the variables and then use PCP on the variates of group means. If the mean of the group means is also calculated, the units can be centred about this overall mean; now the loadings matrix from the group-mean PCP (MPCP) can be applied to the centred variables to obtain scores for the units. These can be plotted on the principal component plots of the group means to show the scatter of the units from each group about their mean.

The above analysis may well be successful. The MPCP analysis will choose directions, in the space of the original variables, that best separate the group means. In other words, these directions will maximize the between-group variation. Unfortunately, these directions may also be those that exhibit a large variation within each group. In that case the means may be widely separated, but the scatter of points for the units about each mean will also be quite large.

Canonical variate analysis (CVA) provides a compromise. It obtains a set of directions such that the ratio of *between-group variability* to *within-group variability* exhibited in each direction is maximized. To explain how this happens we need to concentrate on various sums-of-squares-and-products (SSP) matrices. For PCP the overall, or total, SSP matrix, T say, is used. When the units are grouped there is an SSP matrix within each group; that is, an SSP matrix formed from the units in each group separately. With g groups there will be g such matrices, $W_1, W_2 \ldots W_g$. There is also an SSP matrix between the groups—B. Now T can be decomposed into constituent parts

$$T = B + W_1 + W_2 + \ldots + W_g$$

and, by pooling the within-group SSP matrices,

$$T = B + W$$

As we explained in Section 5.4, a direction is simply a linear combination of the original variables and so can be given as a vector of loadings for the original variables. If x is such a vector, the between-group variation exhibited in the direction defined by x is the matrix product $x'Bx$. Similarly, the within-group variation in that direction is $x'Wx$. Canonical variate analysis finds directions to maximize the ratio $(x'Bx)/(x'Wx)$, whereas the group-mean PCP analysis mentioned above is essentially a PCP analysis of B.

In pooling the within-group SSP matrices, it is assumed that, at least to some extent, these matrices have a similar covariance structure. This is exactly analogous to the pooling of within-group variances when doing a one-way analysis of variance. As in that situation, there are statistical tests available to check the assumption; for example see *Multivariate analysis* (Mardia *et al.* 1979).

Using the life expectancy data from above, the following Genstat statements set up the factors Region, to indicate a grouping of the countries by geographical regions, and Global, to give the "continental" grouping.

```
FACTOR [LABELS = !T('N Africa','Indian Ocean','N America', \
    'C America','Carribean','S America')] Region; \
    VALUES = !((1)2,(2)4,1,3,4,5,4,3,5,(4)2,5,3,(4)2,5,3,(6)4)
FACTOR [LABELS = !T(Africas,Americas)] Global; \
    VALUES = !((1)7,(2)18)
```

We shall use canonical variate analysis first to examine differences between the two global groups, the Africas and the Americas. The directive in Genstat for canonical variate analysis is CVA. It requires as input details of the between and within-group SSPM structures, B and W. These are obtained by first declaring a within-group SSPM structure, and then using FSSPM to form it; FSSPM will also store the group means, and the number of units in each group, from which the matrix B can be calculated by the CVA directive. Note that SSPM will know that W is to be a within-group SSP matrix because of the grouping factor, Global, supplied as the setting of the GROUPS option.

```
SSPM [TERMS = Vars[]; GROUPS = Global] W
FSSPM W
```

Now the CVA directive can be used. The PRINT option is used to request printing of the latent roots and trace, the latent vectors, and the coordinates of the group means in the canonical variate space.

```
  38   CVA [PRINT=roots,loadings,means] W

38.............................................................................

*****  Canonical variate analysis  *****

***  Latent Roots  ***

                    1
                2.523

***  Percentage variation  ***

                    1
               100.00

***  Trace  ***

     2.523

***  Latent Vectors (Loadings)  ***

                         1
            1        21.32
            2       -18.45
            3       -12.94
            4       -10.92
            5       -18.22
            6       -56.15
            7        64.66
            8        11.11
```

```
***  Canonical Variate Means  ***

                       1
          1        2.443
          2       -0.950

***  Adjustment terms  ***

                       1
          1       -44.11
```

The first point to note is that only the first latent root is non-zero. This is because there are only two groups, so only one direction is needed to distinguish between them. Thus there is only one latent vector. Unlike those from principal components analysis, the latent vector does not have sum of squares equal to 1. Remember that CVA is trying to find directions, denoted by x, to maximize the ratio $x'Bx/x'Wx$; clearly x can be multiplied by any (non-zero) constant without altering the ratio, or its maximum. It is usually most convenient to scale x so that $x'Wx = 1$, which is what the CVA directive does. The latent root is the value of the maximized ratio, and so is $x'Bx$, the between-group variation in the direction determined by x.

The loadings give a linear combination of the original variables, from which the scores can be calculated. With CVA, unlike PCP, these are not centred about their means; the centring is achieved by applying an adjustment to the calculated score. The value of the adjustment is printed below the latent vector, and the scores can be calculated as:

$$z_i = 21.3 \times x_{i1} - 18.4 \times x_{i2} + \ldots + 11.1 \times x_{i8} - 44.1$$

where x_{ij} is the data value for the ith country and jth variable, and z_i is the score for the ith country. The adjustment is calculated so that the mean of all the scores for the countries is zero. The scores for the two groups are the *canonical variate means*, the final section of the CVA output above. These can be calculated as the means of the scores for the countries in each group, or equivalently as scores, using the formula above, from the group means of the original data values.

The scores for the countries can be calculated after the analysis. To do this requires the latent vector(s) from the analysis, so the LRV structure CVload is declared before the analysis, and is specified on Line 40 to save the latent root(s) and vector(s). The scores for the countries are obtained by first putting the data values into a matrix (Line 44), and then applying the canonical variate loadings to that matrix (Line 45). The resulting vector of scores then needs to be adjusted, in the same way that an adjustment is applied to the scores for the group means: this simply involves centring the scores for the countries, so that they have zero mean (Line 46). The following Genstat statements will repeat the analysis, and plot the scores.

```
39  LRV [ROWS=Vars; COLUMNS=1] CVload
40  CVA W; LRV=CVload
41  MATRIX [ROWS=Country; COLUMNS=Vars] Data
42  VARIATE Scorevec,Groupno
43  CALCULATE Groupno = Global
44  & Data$[*; 1...8] = Vars[]
45  & Scorevec = Data *+ CVload[1]
46  & Scorevec = Scorevec - MEAN(Scorevec)
47  GRAPH [NROWS=11] Groupno; Scorevec; SYMBOLS=Code
```

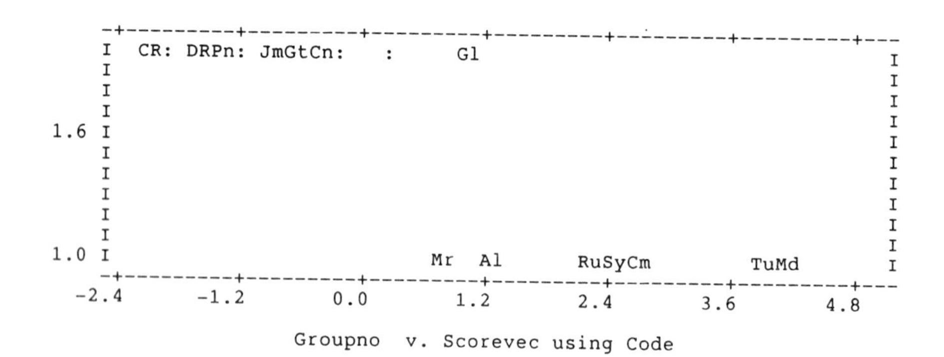

```
     -+---------+---------+---------+---------+---------+---------+---
      I  CR: DRPn: JmGtCn:    :      Gl                                I
      I                                                                I
      I                                                                I
      I                                                                I
  1.6 I                                                                I
      I                                                                I
      I                                                                I
      I                                                                I
      I                                                                I
      I                                                                I
  1.0 I                     Mr   Al        RuSyCm         TuMd         I
     -+---------+---------+---------+---------+---------+---------+---
    -2.4      -1.2       0.0       1.2       2.4       3.6       4.8

                    Groupno  v. Scorevec using Code
```

In the graph above, the y-axis is being used to separate the points for the two groups: those for the "African" countries occur at the bottom of the graph, and those for the Americas are at the top. There is a little bit of overlap, involving the points for Mauritius and Guatemala, but in general the separation of the two groups is quite good. It is fairly easy to see which variables are determining this separation: from the loadings it seems likely that the American countries have higher values for the female variables at ages 50 and 75, and also for males at birth, but probably lower values for the other variables. However, it is difficult to attach any interpretation to this.

Having investigated differences between the two global groups, we shall move on to the analysis between the six regions. Unlike the scores for the countries, which need to be calculated after the analysis, the group-mean scores for the regions can be saved from the analysis. Since there are six regions, the analysis will produce results in (at most) five dimensions; however, it is necessary to save the scores only for the number of dimensions that are required afterwards, so the matrix Rgnscore has been declared with two columns. The following Genstat statements will do the analysis and give the output shown below.

```
48  SSPM [Vars[]; GROUPS=Region] WRegion
49  FSSPM WRegion
50  MATRIX [Region; 2] Rgnscore
51  CVA [PRINT=roots,loadings,means,distances] WRegion; SCORES=Rgnscore
```

```
51.....................................................................

*****  Canonical variate analysis  *****

***  Latent Roots  ***

                   1            2            3            4            5
               4.236        1.870        0.931        0.329        0.249

***  Percentage variation  ***

                   1            2            3            4            5
               55.63        24.56        12.22         4.32         3.27

***  Trace  ***

    7.615

***  Latent Vectors (Loadings)  ***

                     1            2            3            4            5
        1       -47.95       -74.71       144.96       -59.70       -18.38
        2       -45.89        74.70         1.40       -77.82       -44.84
        3         7.24       -21.88       -29.56        58.60        11.94
        4       -15.25         7.26        -1.20        -2.42        -3.09
        5        59.01        23.84      -141.42        58.48       -35.06
        6       -83.59        22.89        36.28        39.78       177.70
        7       116.14       -24.43       -43.63       -28.95       -49.14
        8        -7.56         3.18        26.20        -0.61       -12.28

***  Canonical Variate Means  ***

                     1            2            3            4            5
        1       2.2712       2.6904       0.0279      -0.2989      -0.2307
        2       2.5242      -1.2501       1.0917       0.3463       0.0361
        3       0.9712      -1.0945      -1.6235      -0.1232      -0.2488
        4      -2.1548       0.1161       0.3391       0.2614      -0.4921
        5      -1.1233      -0.4996       0.4592      -0.9641       0.3553
        6      -0.8432       0.6524      -0.4570       0.5733       0.7686

***  Adjustment terms  ***

                     1            2            3            4            5
        1       -40.85        16.96       -11.92       -25.46        59.94

***  Inter-group distances  ***

   1        0.000
   2        4.149        0.000
   3        4.333        3.180        0.000
   4        5.167        4.961        3.911        0.000
   5        4.761        4.011        3.186        1.918        0.000
   6        3.981        4.236        3.037        2.081        2.186        0.000
                1            2            3            4            5            6
```

Of the five latent roots, only the first two are greater than 1. Since these are the ratios of between-group to within-group variation in each dimension of the solution, this indicates that dimensions 3, 4, and 5 all exhibit more within-group variation than between-group variation. There are five vectors of loadings, corresponding to the five latent roots, and each has an adjustment term, printed below the canonical variate means for the groups.

Printed below the adjustments is a symmetric matrix of Mahalanobis's distances, requested by the setting 'distances' of the PRINT option of the CVA directive.

These are the distances between the group means in the space of the canonical variate analysis solution; that is, they are the distances between the canonical variate means. These distances are also the values given by the following formula. If y_i is a vector with the mean values, for the countries in the ith group, of the eight variables, then the Mahalanobis squared-distance between the ith and jth groups is

$$d^2_{i,j} = (y_i - y_j)' \, W^{-1} \, (y_i - y_j).$$

Thus Mahalanobis distance is adjusted for within-group variation. From the printed values it can be seen that the truly African countries are dissimilar to all the other groups (with the exception, to some extent, of the Indian Ocean countries). The most similar groups of countries are those from the Caribbean and Central America, which is not very surprising.

A plot of the group means in the first two dimensions is produced by the following Genstat statements:

```
52  VARIATE [NVALUES=Region] Rscore[1,2]
53  CALCULATE Rscore[] = Rgnscore$[*; 1,2]
54  GRAPH [EQUAL=scale; NROWS=37; NCOLUMNS=61] Rscore[2]; Rscore[1]; \
55     SYMBOLS=!T(Af,IO,NA,CA,Cb,SA)
```

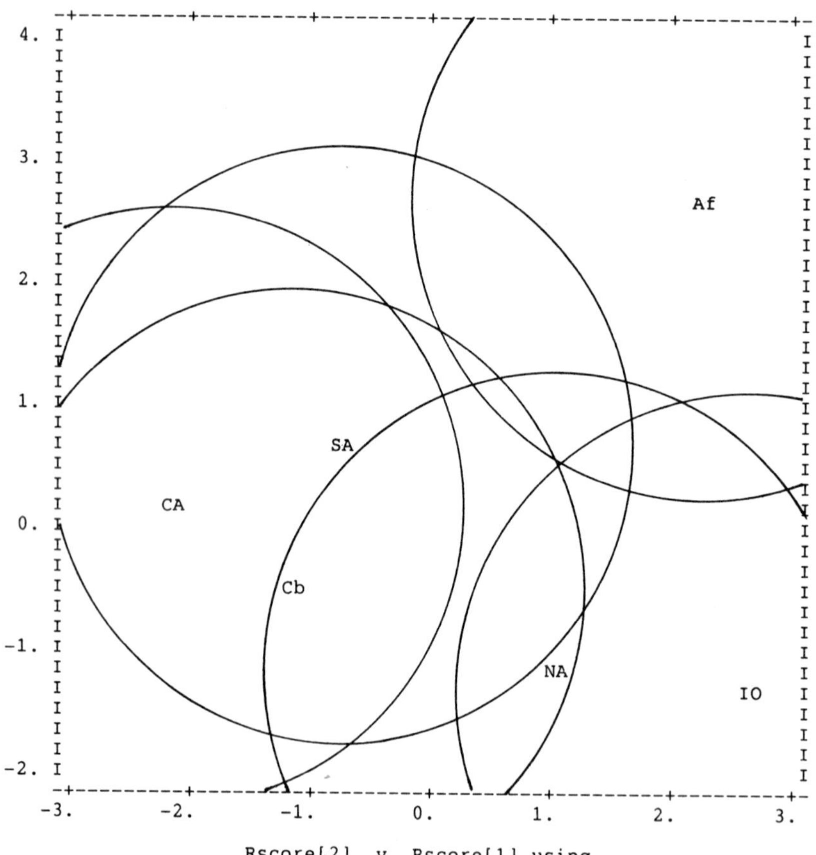

The relative remoteness of the point for Africa is apparent; also the closeness of the points for the Caribbean and Central American groups can be seen. The proximity of the point for South America to that for the Caribbean is slightly misleading, since their Mahalanobis distance is 2.2.

It would be quite feasible to plot the scores for the individual countries on the same graph. Calculating such scores would be similar to the calculations on Lines 45 to 47 above. However, an extra step would be needed here, because more than one column of scores is to be formed: the result of post-multiplying the matrix Data by the canonical variate loadings would have to be stored in a matrix, and its columns extracted into variates, as has been done already in Section 5.3. However, we have adopted a different device to show within-group scatter. It was noted above that the latent vectors are scaled so that $x'Wx = 1$. In the general case, where there are several latent vectors, forming the matrix X, the scaling is such that $X'WX = I$, the identity matrix (here of order 5, the number of latent vectors). This means that within the canonical variate space within-group variation is unity in every direction. So, if it is assumed that the data have a multivariate Normal distribution with equal within-group covariance matrices, a 95% "confidence" circle can be drawn around each group mean. The radius of this circle is given by the square-root of the 95% point of a chi-squared variable; the number of degrees of freedom is 2 because the plot is two-dimensional. The appropriate value is 2.45, and we have drawn such circles on the plot above. These circles should include 95% of the points for the countries in each region. It is quite common to find graphs from CVA with these confidence circles drawn on them. Another common device is to draw 95% confidence circles for the position of the group mean: these are simply circles with radii $2.45/\sqrt{n_i}$, where n_i is the size of the ith group. Almost as common, unfortunately, is the practice of drawing circles on the graph, without explaining what they are!

5.7 Exercises

5(1) Mass-spectrometry was used to determine the amounts, in parts per million, of five chemical elements in 26 insects. The elements are Potassium (K), Chlorine (Cl), Sulphur (S), Silicon (Si), and Magnesium (Mg). The data are given below.

21115	1276	5854	1046	2877
13034	915	5430	1257	3251
13877	1140	5573	1452	3081
11867	1017	6218	537	2860
15020	878	5528	540	2901
15384	1099	6452	570	2759
10666	955	6550	588	3102
13836	1105	6236	257	2926
12605	985	5302	196	2559
15139	476	4220	872	2886

22797	562	4503	904	3136
16329	522	5290	548	3405
13132	465	5269	800	4049
21183	531	4638	498	3161
15505	477	4481	455	2925
14266	595	5080	713	3163
21526	526	4875	529	3411
31465	857	5931	532	3942
13047	741	5527	777	2932
12747	649	4644	555	3082
27748	967	5518	578	3295
18456	895	4563	665	3099
15006	568	5784	647	3899
16633	819	5106	462	3555
27053	1310	5848	790	4111
21051	921	4737	673	3090

(Data from those used by Sherlock, Bowden, and Digby, 1985).

Write a Genstat program to analyse these data using principal components analysis. Produce a scatter plot of the scores for the first two principal components.

5(2) Using the data from Exercise 5(1), write a program to take a logarithmic transformation of the data and form a biplot of the data. Compare this with a biplot formed from the principal components analysis of the correlations of the untransformed data.

5(3) The 26 insects for Exercise 5(1) are in three groups: units 1 to 9, units 10 to 17, and units 18 to 26. Write a program to take a logarithmic transformation of the data and analyse the three groups using canonical variate analysis. Produce a scatter plot of the canonical variate scores for the units and group means.

6 Analyses of distances

6.1 Introduction

In Chapter 5 we described two multivariate methods which are based on sums of squares and products. In this chapter we shall look at two methods based on the idea of distances between different things.

To introduce the methods we consider some data from Loxdale *et al.* (1985). The data are derived from field samples of about 1000 aphids (of a particular type) taken at six different sites in Britain. Using electrophoresis the frequency of different alleles for the six samples were obtained, and this provided a genetic comparison of the samples. The data shown below are measures of genetic distance among the six samples, so that the values indicate how different the samples are in terms of their genetic make-up. (For convenience the values below are 1000 times the published data.)

	East Craigs	Long Ashton	Rothamsted	Henlow	Bayfordbury	Norwich
East Craigs	0					
Long Ashton	90	0				
Rothamsted	105	70	0			
Henlow	40	46	66	0		
Bayfordbury	82	76	102	34	0	
Norwich	21	72	81	19	50	0

For example, the genetic distance between the samples from Long Ashton (near Bristol) and Rothamsted (near St Albans) is 70. A natural question is: how might we analyse these data and display the distances?

6.2 Principal coordinate analysis

This method is intended precisely for handling data such as the genetic distances. The method finds (or, at least, tries to find) a set of coordinates of points representing the six field samples so that the distances between the points are the same as the genetic distances. Thus two samples that are genetically similar will be represented by two points that are close together.

In Genstat the acronym PCO is used for Principal CoOrdinate analysis to avoid confusion with Principal ComPonents analysis. The algebraic details of PCO need not concern you, except in two aspects. The first is that it is easier to work with squared distances rather than distances; the second is that the formulation in terms of similarities, rather than distances, is sometimes convenient. In fact the analysis produces coordinates of points such that the squared distance between two points is approximately

$$d^2_{ij} = a_{ii} + a_{jj} - 2 \times a_{ij}$$

where a_{ij} are the values input to the analysis. If the original data are considered as distances, d_{ij}, we need to transform them to

$$a_{ij} = - d^2_{ij}/2$$

(since $a_{ii} = a_{jj} = 0$) before the analysis. With this in mind a Genstat program can be written to read the data into a symmetric matrix, transform them, and analyse them with the PCO directive. The output from such a program is shown below.

```
   1  TEXT Places; !T('E Craigs','Long Ashton','Rothamsted','Henlow', \
   2                 'Bayfordbury','Norwich')
   3  SYMMETRIC [ROWS=Places] Gendists,Transdis
   4  READ Gendists

    Identifier   Minimum       Mean   Maximum     Values   Missing
      Gendists      0.00      45.43    105.00         21         0
   6  CALCULATE Transdis = - Gendists * Gendists / 2
   7  PCO [PRINT=roots,scores] Transdis


7...................................................................


*****  Principal coordinates analysis  *****

***   Latent Roots  ***

                     1            2            3            4            5
                  7160         3716         1894            0          -79
                     6
                  -824

***   Percentage variation  ***

                     1            2            3            4            5
                 60.33        31.31        15.96         0.00        -0.66
                     6
                 -6.94

***   Trace  ***

      11867

*  Some roots are negative - non-Euclidean distance matrix  *


***   Latent vectors (coordinates)  ***

                           1           2           3
               1      -41.37      -31.96       10.47
               2       31.27       17.59       32.97
               3       59.02      -20.47      -19.00
               4       -5.10        3.35       -0.25
               5      -21.12       42.75      -16.78
               6      -22.70      -11.27       -7.41

*  Vectors corresponding to zero or negative roots are not printed  *
```

The analysis is based on a spectral decomposition that gives *latent roots* and vectors. As with principal components analysis, the latent roots can be interpreted as amounts of variation contained within each of several dimensions. Also the amount of variation in each dimension is the maximum possible, subject to the dimension being orthogonal to each of its predecessors. With PCO the variation is considered in terms of the (negative) squared distances input to the directive; for example, it can be seen from the output that the first two dimensions explain nearly 92% of the total squared distance.

However, with principal coordinate analysis there can be *negative latent roots*, as in this example where the last two roots are −79 and −824. There is a simple explanation for this somewhat surprising feature. It means that it is impossible to find a set of points, in any number of dimensions, which exactly represent the initial set of distances. The best that can be done with these data is to represent the field samples by points in three dimensions. The first three percentage-variance values sum to 107.6%, so the three-dimensional solution overfits the distances by 7.6%. The last two roots correspond to *imaginary dimensions* in which some squared distances are negative. The message preceding the percentages is a warning of negative roots. It is not really needed here because the analysis has printed all the roots, but it is useful if only some of the roots are being printed.

Another feature of PCO is that there will always be one zero root—here it is the fourth—and there may be more than one. This is because at most $(n - 1)$ dimensions are needed to fit n points, if it can be done at all. For example, two points can be fitted onto a line, and three points into a plane to form a triangle. Each extra point may introduce one dimension, but no more than one.

The coordinates of the points for the field samples are printed after the latent roots. In fact they are a set of latent vectors, but scaled so that the variation in each dimension, that is the sum of squares of each column, is the same as the corresponding latent root. The imaginary dimensions corresponding to the negative roots are not printed because it is impossible to have a column of real values whose sum of squares is negative! Nor is the column of zeroes corresponding to the fourth root printed. Because each dimension contains a maximum amount of variation and the dimensions are orthogonal they are called *principal axes*, and the points are said to be referred to their principal axes.

It is useful to plot the points for the six field sample sites. To do this the coordinates of the points must be saved: since these are latent vectors an LRV structure must be declared first; this is then used in the PCO statement.

```
LRV [ROWS = Places; COLUMNS = 2] Lrv
PCO Transdis; LRV = Lrv
```

Note that only the first two dimensions, and correspondingly only the first two latent roots, will be saved from the analysis. Also the PRINT option of PCO has not been used, so there will be no printed output from the analysis. Having saved the coordinates, the following statements will plot the points in the first two dimensions to produce the graph shown below.

```
10  VARIATE [NVALUES=Places] Coords[1,2]
11  CALCULATE Coords[] = Lrv[1]$[*; 1,2]
12  GRAPH [EQUAL=scale; NROWS=37; NCOLUMNS=61] Coords[2]; Coords[1]; \
13     SYMBOLS=Places
```

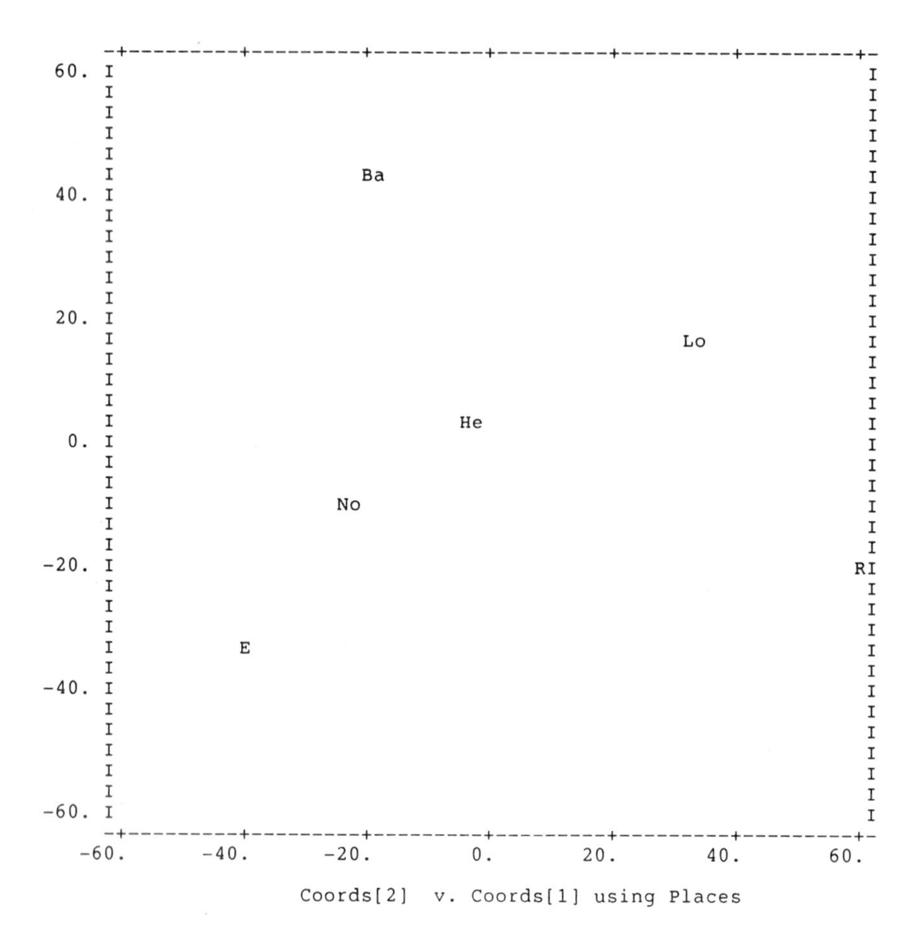

The graph shows that the three mutually most distant points are for Bayfordbury, East Craigs, and Rothamsted, in accord with the data. The genetic distance between Bayfordbury and Rothamsted is somewhat surprising, since they are fairly close geographically.

As well as the PRINT option settings 'scores' and 'roots', other settings can be used to request printed output. It is useful sometimes to see if any points are very far from the rest: if the setting 'centroid' is used PCO will print the distances of each point from the centroid of all the points, so that relatively large values will indicate outliers. These *centroid distances* are calculated as if an exact solution were possible. The second option of PCO is NROOTS and can be set so that results corresponding to only some of the dimensions are printed. If the NROOTS option has been used, so that the points are fitted only approximately, the residual distances can be printed using the PRINT option setting 'residuals'. The third option of PCO specifies whether results are to be printed corresponding to the largest or smallest latent roots: by default it is the former; the option setting SMALLEST = yes specifies the latter.

The squared distances of the points from their centroid and the residuals can also be saved from the analysis as a diagonal matrix and a matrix with one column, respectively. They must be declared in advance of the PCO statement; the parameters for them are CENTROID and RESIDUALS. It is generally more useful for analyses that might follow principal coordinate analysis to work with squared distances of the points from their centroid; so the values saved using the CENTROID parameter are squared distances, as opposed to the printed centroid distances which are not squared.

6.3 The relationship between principal components and principal coordinate analysis

In the last example the original data were in the form of distances. Although distance data occur from time to time, it is far more common to have data in the form of a units-by-variates data matrix. In that case the data can be considered as giving the coordinates of points for the units, with each variate representing one dimension. We showed this interpretation in Chapter 5, Section 5.4 with the PCP example. The coordinates could be used to calculate distances among the points and these used as input to PCO. However, the results of such an analysis would be identical with the PCP results, provided that distances had been calculated in the obvious way: that is, the minimum distance in a straight line between each pair of points. This is because the two analyses are both trying to represent the points, and thus the distances among them, as well as possible.

Because of the equivalence of the two analyses it is possible in Genstat to use the PCO directive for principal components analysis. This is done by replacing the symmetric matrix of data by a pointer containing a list of variates. The PCO directive will calculate the symmetric matrix of distances among the points and then analyse that. Of relevance to the genetic distance data is the ability to save the calculated distance matrix by using the keyword DISTANCES and supplying a symmetric matrix to hold the results. The PCO directive will also print the distance matrix if the PRINT option setting 'distances' is used. If the coordinates of the two-dimensional solution are input to PCO the distances displayed in the graph can be saved. Now the symmetric matrix of residual, or not fitted, distances can easily be calculated. Note that these residuals are not the same as the printed residuals obtained by including 'residuals' in the PRINT option setting: this refers to residual distances for the points, not between all pairs of points. The output from the Genstat statements to do this is shown below.

```
14   SYMMETRIC [ROWS=Places] Fitdists,Resdists
15   PCO Coords; DISTANCES=Fitdists
16   CALCULATE Resdists = Gendists - Fitdists
17   PRINT Gendists,Fitdists,Resdists

                 Gendists

    E Craigs         0.00
 Long Ashton        90.00        0.00
  Rothamsted       105.00       70.00        0.00
      Henlow        40.00       46.00       66.00        0.00
 Bayfordbury        82.00       76.00      102.00       34.00        0.00
     Norwich        21.00       72.00       81.00       19.00       50.00

                 E Craigs Long Ashton  Rothamsted      Henlow Bayfordbury


     Norwich         0.00

                  Norwich


                 Fitdists

    E Craigs         0.00
 Long Ashton        87.93        0.00
  Rothamsted       101.05       47.11        0.00
      Henlow        50.62       39.06       68.40        0.00
 Bayfordbury        77.41       58.12      102.08       42.53        0.00
     Norwich        27.87       61.20       82.24       22.88       54.04

                 E Craigs Long Ashton  Rothamsted      Henlow Bayfordbury


     Norwich         0.00

                  Norwich


                 Resdists

    E Craigs        0.000
 Long Ashton        2.070       0.000
  Rothamsted        3.955      22.895       0.000
      Henlow      -10.618       6.941      -2.403       0.000
 Bayfordbury        4.593      17.884      -0.075      -8.533       0.000
     Norwich       -6.872      10.801      -1.236      -3.876      -4.040

                 E Craigs Long Ashton  Rothamsted      Henlow Bayfordbury


     Norwich        0.000

                  Norwich
```

The matrix of residual distances shows that certain distances have been overfitted; for example, the distance on the graph between East Craigs and Henlow is 50.6 but the genetic distance is only 40. Adding dimensions cannot reduce the distance between points and this is where the imaginary dimensions come in. Reference to the original PCO output shows that the fitted squared distance between East Craigs and Henlow is increased in the third real dimension by $(10.47 + 0.25)^2$, which makes the fitted squared distance in the three-dimensional solution 2675. The true squared distance is 1600, so the discrepancy in squared distance of -1075 must be fitted in the two imaginary dimensions. Sometimes these imaginary distances can be informative in revealing the non-Euclidean features of the data. The pattern of negative values in the matrix of residual distances suggests that it is the data for the last four sites that are causing the non-Euclideanarity.

Although PCO will mimic PCP it can be used more generally when the data are of the units-by-variates form. When a set of variates is input to PCO the directive calculates distances among the points as straight-line distances. These are called *Pythagorean distances* and they are formed by the natural extension of Pythagoras' theorem:

$$d_{ij}^2 = \sum_k (x_{ik} - x_{jk})^2$$

However, with some sets of variates this formulation of distance is not desirable. There are many other methods of calculating distance, which are usually called *metrics*; for example the Manhattan, or city-block, metric:

$$d_{ij}^2 = \sum_k |x_{ik} - x_{jk}|$$

In these situations it is necessary to form the distance matrix separately and then analyse it with PCO. In Genstat this is easiest to do by forming a similarity matrix, rather than a distance matrix, with the FSIMILARITY directive which is described in Chapter 7.

6.4 Procrustes rotation

The experimenter who obtained the genetic distance data of the last example believes that genetic distance increases with geographical distance. If the genetic distances were actually the same as the geographical distances then one might expect the principal coordinate analysis to produce a map in which the points for the six field sites corresponded to their geographical location. One of the sites (East Craigs) is near Edinburgh; another (Long Ashton) is near Bristol; the other four are in the East of England. It is clear that the graph produced above does not correspond to a geographical map; however, this need not be because the genetic distances are different from the geographical distances. In the remainder of this chapter we shall explore one approach to the problem of linking the PCO map to a geographical map.

As with many sets of data, there is no clear-cut choice of the "right" analysis; the ideas described below are simply one way of tackling the problem.

The reason why the graph from PCO does not match one's preconceived geographical ideas (which, for example, would require that the point for East Craigs be near to the top of the graph) may be because the coordinates from PCO are not unique. For example, any constant could be added to all the values in any column of the matrix of coordinates without altering the inter-point distances. Such a transformation would merely shift the centroid of the six points from the origin, which is where PCO always locates the centroid. Similarly, any orthogonal rotation of the principal axes will change the positions of the points, but will leave the inter-point distances unchanged. For example, if you turn the graph clockwise through about 120 degrees you will see the point for East Craigs near the top, as required; although the other points still do not match the geography.

So here is a set of six points which can be considered as fixed, namely the geographical locations. There is also another configuration of six points which are determined only by the distances between them. These can be moved, either by a change of origin or by a rotation of the axes, without altering their interrelationships. Procrustes rotation is the name given to the method that finds the best change of origin, followed by a rotation of the axes, so that the latter configuration of points matches the former configuration as closely as possible. The phrase "as closely as possible" means that the sum of squared distances between the fixed points and their corresponding transformed points is minimized.

The term "Procrustes" rotation has been coined after a rather unsavoury character from Greek mythology. Procrustes was an innkeeper who had only one bed for guests, and had the practice of spiking his guests' drinks. Having got the unsuspecting fellow drunk, he made the guest fit the bed exactly, either by lopping off limbs or by stretching the guest on a rack. Although these activities do not match exactly the ideas of changing the origin or rotating the axes, the term has stuck.

An easy way to obtain coordinates for the six field sites is to take Easting (x-coordinate) and Northing (y-coordinate) values from Ordnance Survey maps. Relative to an arbitrary origin (somewhere to the South of Devon) the following have been obtained; they are expressed in kilometres.

	Easting	Northing
East Craigs	320	673
Long Ashton	354	171
Rothamsted	514	204
Bayfordbury	531	208
Norwich	623	308
Henlow	518	238

To use Procrustes rotation in Genstat, matrices holding the two sets of coordinates, those from the principal coordinate analysis and those given above, are needed. The following Genstat statements will obtain these.

```
 18  MATRIX [ROWS=Places; COLUMNS=2] OSlocs
 19  READ OSlocs

   Identifier   Minimum      Mean   Maximum    Values   Missing
       OSlocs     171.0     388.5     673.0        12         0
 21  LRV [ROWS=Places; COLUMNS=3] Lrv3; VECTORS=PCOlocs
 22  PCO Transdis; LRV=Lrv3
```

The coordinates for the six sites are read in to the matrix OSlocs. The PCO results are stored in the LRV structure Lrv3; although only two dimensions from the principal coordinate analysis were plotted above, all three (real) dimensions will be used for the Procrustes rotation. Although the matrix of coordinates can be referred to with the identifier Lrv3[1], it is more convenient below to use the identifier PCOlocs. We have used the VECTORS parameter of the LRV directive to associate the identifier PCOlocs with the matrix of vectors that is part of the LRV structure. In fact all the components structures of Genstat's compound structures—SSPMs, LRVs, and TSMs—can be referred to by identifiers that you choose.

In Genstat, Procrustes rotation is provided by the ROTATE directive. It requires as input the two matrices of coordinates, given as separate lists. The first matrix gives the locations of the fixed points, so here it is the matrix OSlocs. The second matrix contains the coordinates that are to be transformed. In common with all the multivariate directives, ROTATE will not print any results unless specifically asked to. To get the coordinates of the two sets of points printed the setting 'coordinates' of the PRINT option of ROTATE must be set. The following Genstat statement will do the analysis and print both configurations of points. The output is shown below.

```
  23  ROTATE [PRINT=coordinates] OSlocs; PCOlocs

 23.....................................................................

 *****  Procrustes rotation  *****

 ***  Fixed Configuration  ***

                          1            2            3
             1      -0.3169       0.7538       0.0000
             2      -0.2481      -0.2616       0.0000
             3       0.0755      -0.1949       0.0000
             4       0.1099      -0.1868       0.0000
             5       0.2960       0.0155       0.0000
             6       0.0836      -0.1261       0.0000

 ***  Fitted Configuration  ***

                          1            2            3
             1      -0.1610       0.4421      -0.0351
             2      -0.1964      -0.3226       0.2080
             3      -0.0824      -0.4132      -0.3954
             4       0.0284       0.0310       0.0340
             5       0.3759       0.0375       0.2395
             6       0.0354       0.2252      -0.0510
```

The first thing to notice is that the fixed configuration of geographical points has been changed. By default ROTATE arranges that the centroids of the two sets of points are at the origin, which merely involves subtracting the mean x- and y-coordinates from the Ordnance Survey grid references (since the PCO results are already centred at the origin). The other transformation, which has been done by default, is to rescale the axes for each configuration of points so that the total sum of squares for each configuration is equal to one. However, these two transformations have not really altered the geographical locations; a plot of the fixed configuration coordinates would still show the sites in the expected places, but the scales on the axes would have changed.

The centring of both sets of points to the origin is particularly convenient. The algebraic definition of Procrustes rotation requires that the centroids of each configuration be the same for the "best" fitting solution. Also, the rotation that is required to get the best solution is done about this centroid, and this is easiest to do if the centroid is at the origin. However, the rescaling of the axes is less convenient, because the scales are now arbitrary. It would be better if both sets of points were centred, but the axes were not rescaled. This would change the arbitrary origin of the geographical map from somewhere in the English Channel to somewhere near Bedford, but would retain the kilometre scaling of the axes.

To obtain the centring, but not the rescaling, we need to use the STANDARDIZE option of ROTATE. This has two possible settings, 'centre' and 'normalize'; the default setting is STANDARDIZE = centre,normalize so this option should be set to 'centre' for this purpose.

```
  24  ROTATE [PRINT=coordinates; STANDARDIZE=centre] OSlocs; PCOlocs

24..........................................................................

*****  Procrustes rotation  *****

***  Fixed Configuration  ***

                       1            2            3
           1     -156.67       372.67         0.00
           2     -122.67      -129.33         0.00
           3       37.33       -96.33         0.00
           4       54.33       -92.33         0.00
           5      146.33         7.67         0.00
           6       41.33       -62.33         0.00

***  Fitted Configuration  ***

                       1            2            3
           1      -18.20        49.96        -3.96
           2      -22.19       -36.46        23.51
           3       -9.31       -46.70       -44.68
           4        3.21         3.50         3.84
           5       42.48         4.24        27.06
           6        4.01        25.45        -5.77
```

The fixed configuration of points is now more familiar; for example, the Northerly position of East Craigs is shown by the second coordinate of 373. Inspection of the fitted configuration shows that the coordinate values have (mostly) the same sign as their counterparts in the fixed set of points. However, all the values are too small. This is caused by a difference in the two scales of measurement. While the geographical locations are expressed in kilometres, the genetic distances are not, and are not commensurate with the geographic distances.

Clearly, the coordinates produced by the PCO analysis must be expanded. However, this must be done isotropically; that is, each dimension of the PCO results must be expanded by the same amount. If the required multiple was known this could be achieved by multiplying the genetic distances before transforming and analysing them with PCO. Since the correct value for the multiplier is unknown (although one might guess that it is about 5), the ROTATE directive must be allowed to do the expansion itself. This expansion is called "scaling" (not to be confused with the initial scaling of the configurations, which has been suppressed). To allow it the second option of ROTATE must be set: SCALING = yes.

Some additional PRINT option settings can also be set in ROTATE. The setting 'rotations' requests that the orthogonal rotation matrix be printed; the settings 'residuals' and 'sums' stand for residuals and sums of squares, respectively. To save the coordinates of the (centred) fixed and the fitted configurations, matrices must be declared and given in the ROTATE statement. These are separate parameters (the third and fourth) of the directive, with parameter names XOUTPUT and YOUTPUT.

```
  25  MATRIX [ROWS=Places; COLUMNS=3] OSlocs3,Fitlocs
  26  ROTATE [PRINT=coordinates,rotations,residuals,sums; SCALING=yes; \
  27     STANDARDIZE=centre] OSlocs; PCOlocs; XOUTPUT=OSlocs3; YOUTPUT=Fitlocs

27.....................................................................


*****  Procrustes rotation  *****

***  Orthogonal Rotation  ***

                     1           2           3
          1   -0.20881    -0.92849    -0.30708
          2    0.58221    -0.37032     0.72381
          3   -0.78577    -0.02764     0.61790

***  Fixed Configuration  ***

               OSlocs3
                     1           2           3
      Places
    E Craigs   -156.67      372.67        0.00
 Long Ashton   -122.67     -129.33        0.00
  Rothamsted     37.33      -96.33        0.00
      Henlow     54.33      -92.33        0.00
 Bayfordbury    146.33        7.67        0.00
     Norwich     41.33      -62.33        0.00
```

```
***   Fitted Configuration   ***

                    Fitlocs
                          1            2            3
        Places
      E Craigs       -53.77       147.62       -11.71
   Long Ashton       -65.58      -107.74        69.46
    Rothamsted       -27.50      -137.99      -132.04
        Henlow         9.48        10.35        11.35
   Bayfordbury       125.53        12.54        79.97
       Norwich        11.84        75.21       -17.04

***   Residuals   ***

                          1
             1        247.7
             2         92.5
             3        152.9
             4        112.6
             5         82.8
             6        141.7


*** Sums of Squares ***

Fitted Configuration        111508.8906
Residual                    132907.7656
-------------------------------------
Fixed Configuration         244416.6562

***   Least-squares Scaling factor =          2.9550
```

The orthogonal rotation matrix gives loadings of the coordinates of the PCO solution to form the fitted configuration. For example, the first coordinate of the fitted configuration is

$$- 0.21 \times x_{i1} + 0.58 \times x_{i2} - 0.79 \times x_{i3}$$

where x_{ij} is the coordinate of the *i*th point in the *j*th dimension of the PCO results. Interestingly, it is the third dimension of the PCO solution that contributes most to the North-South axis.

The fixed configuration is, as before, expressed on the original kilometre scale. Comparing the two configurations shows some marked discrepancies: the fitted point for East Craigs is too far South; that for Rothamsted is too far West and South. The residuals, printed below the configurations, emphasize these points; for example, the fitted point for East Craigs is nearly 250 kilometres from its true location.

The penultimate output from ROTATE is a summary table showing the sums of squares of each configuration about their common centroid. In this instance, the difference between the two sums of squares for the configurations is actually the sum of squares of the residuals. This is always the case when the option SCALING = yes is set in ROTATE. However, for reasons associated with the algebraic solution of the Procrustes rotation, it is not generally the case when scaling is not allowed.

When scaling is used, as here, the value of the scaling factor is printed along with the sums-of-squares table. For this analysis its value is of no interest; however, when the original scales of measurement are comparable it can be. For example, if inter-town road distances are used as input to PCO the resulting "map" is generally larger than the true map because roads are not straight; now the value of the scaling parameter is a measure of how much the roads wind in the region under consideration.

Having saved the resulting configurations of points from ROTATE, it is straightforward to draw a graph showing the true locations and those from the fitted configuration. To distinguish between the two sets of points the true locations can be represented by upper-case codes and the fitted locations by lower-case codes. The following Genstat output shows the statements used to produce the graph, and the graph itself.

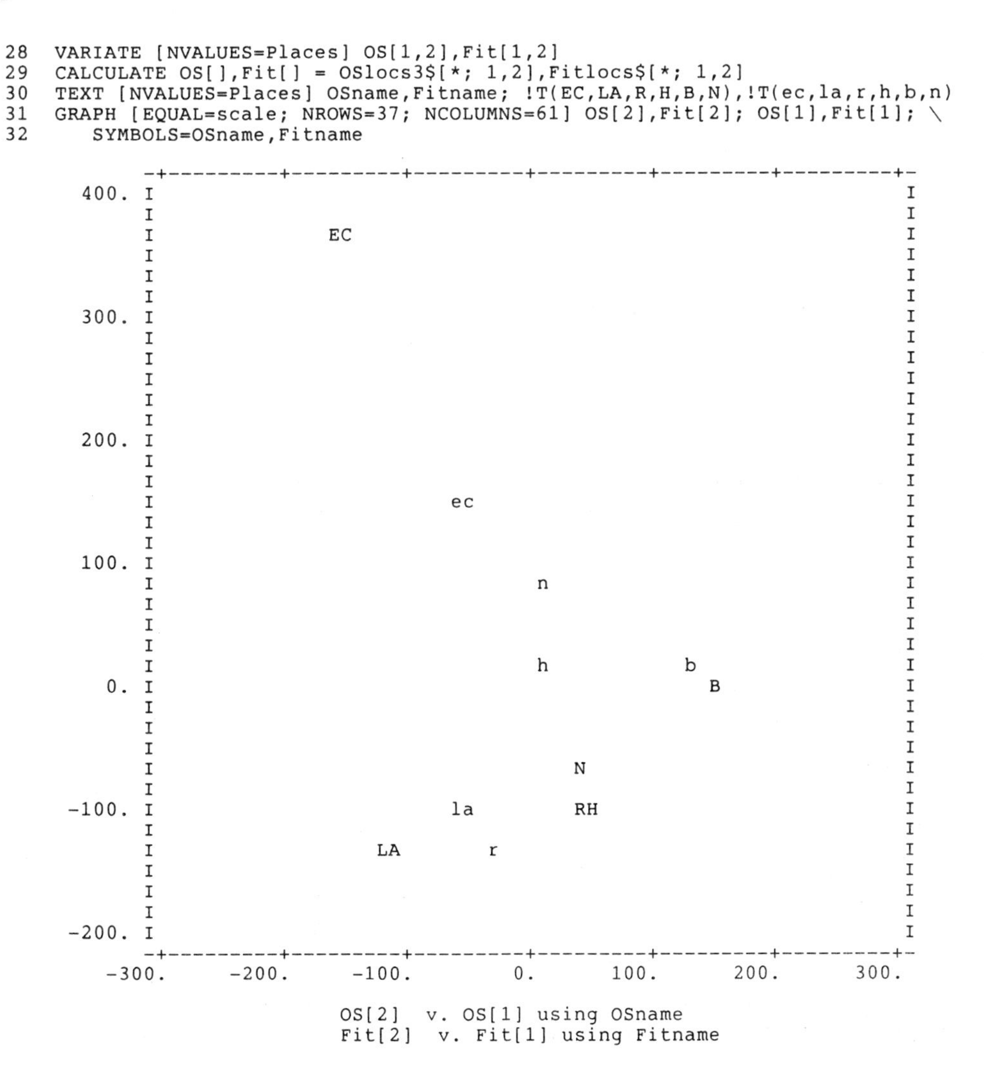

```
28  VARIATE [NVALUES=Places] OS[1,2],Fit[1,2]
29  CALCULATE OS[],Fit[] = OSlocs3$[*; 1,2],Fitlocs$[*; 1,2]
30  TEXT [NVALUES=Places] OSname,Fitname; !T(EC,LA,R,H,B,N),!T(ec,la,r,h,b,n)
31  GRAPH [EQUAL=scale; NROWS=37; NCOLUMNS=61] OS[2],Fit[2]; OS[1],Fit[1]; \
32      SYMBOLS=OSname,Fitname
```

Now the ill-fitting of the point for East Craigs is made apparent. It might be postulated that the data, both genetic and geographic, for this site are responsible for the overall poorness of fit. Geographically, East Craigs is twice as far from any other site as the other sites are from each other. However, the same is not true of the genetic distances; although East Craigs is genetically distant from some of the other sites, it is not in the same 2:1 ratio. Perhaps an analysis for the five English sites would give a better genetic-to-geographic fit—we shall leave that as an exercise for you.

6.5 Exercises

6(1) Write a program to repeat the principal coordinate analysis and Procrustes rotation using the genetic distance data for the five English sites only.

6(2) The actual locations of 16 underground stations in London are given below as (x,y) coordinates. Their locations according to a London Underground map are also given as (x,y) coordinates. (Both sets of coordinates are referred to arbitrary origins; the actual locations are in units of 1/90th mile). Write a program to compare the two sets of coordinates using Procrustes rotation. This should allow you to assess which stations have their actual locations poorly represented on the underground map.

Station	Actual location		From Underground map	
Oxford Circus	73	61	41	36
Bond Street	44	53	24	36
Regent's Park	52	111	41	43
Tottenham Court Road	118	68	60	36
Covent Garden	142	47	81	17
Leicester Square	126	37	60	17
Piccadilly Circus	103	27	47	17
Green Park	71	6	23	17
Marble Arch	10	47	6	36
Baker Street	15	104	24	63
Great Portland Street	64	114	32	60
Warren Street	85	120	60	56
Goodge Street	101	94	60	46
Holborn	158	76	91	34
Aldwych	174	44	78	14
Charing Cross	135	16	60	3

6(3) Using the data from Chapter 5, Exercise 5(1), write a program to take a logarithmic transformation of the data, and then to use the PCO directive for principal components analysis. Analyse the correlations of the untransformed data with principal components analysis, and compare the two sets of scores from the principal components analyses using Procrustes rotation.

7 Cluster analysis

7.1 Introduction

In Chapter 5 we described canonical variate analysis as a method for using a predetermined knowledge of the grouping of a set of units to derive rules for discriminating among the groups. In this chapter we shall look at some methods that attempt to find a natural grouping among units based on data for those units. There are many ways of doing this, of which the most popular are the methods of agglomerative hierarchical classification: often the phrase "cluster analysis" is used for these techniques.

Cluster analysis methods do not use as input a units-by-variates data matrix. Instead they use a *similarity matrix*. This is a symmetric matrix, like the distance matrices introduced in Chapter 6. However, as the name suggests, its values are the similarities of each unit to each other unit. Conventionally, similarities lie in the range zero to one. Self-similarities, that is the similarities of units with themselves, are one. Because the matrix is symmetric, the similarity of the ith unit to the jth unit is the same as the similarity of the jth unit to the ith. Similarity values are often given as percentages in the range 0 to 100.

7.2 Similarity matrices

Sometimes cluster analysis is used to determine whether a set of objects do divide into distinct groups, or whether they come merely from a single population. This is relevant to archaeologists investigating the artifacts recovered from excavation sites: if the artifacts seem to come from distinct groups then it might be considered that the site was used by distinct groups of individuals. We shall look at cluster analyses of a small sample of 28 bronze brooches that are among those recovered from Iron-age graves in the Munsingen cemetery. The brooches look roughly like safety pins; they have a pin, with one end floating, which can close across a bow of typically crescent shape. There are various ornamentations, including a series of coils where the bow and pin are joined; presumably these gave the spring to the pin. The data, from *Mathematics and Computers in Archaeology* (Doran and Hodson 1975), consist of 14 variables. Of these, two are the angles at the front and rear of the bow, another is also an angular measurement, and one is the number of coils. The other 10 are lengths of various parts of the brooch, including its total length. Unfortunately, one variable cannot be measured for five of the brooches, so its value is missing for those; similarly, another variable has one missing value. In this chapter we shall try to find out if there is any natural grouping of these brooches; for example, are they of two, or more, distinct styles. The first thing to do is read the data: the Genstat output is shown below.

```
 1  UNITS [NVALUES=28]
 2  READ Vset[1...14]

  Identifier   Minimum       Mean   Maximum     Values   Missing
     Vset[1]     10.00      29.93     94.00         28         0      Skew
     Vset[2]      7.00      16.64     26.00         28         0
     Vset[3]     1.000      3.000     7.000         28         0
     Vset[4]     6.000      8.286    10.000         28         0
     Vset[5]     5.000      8.000    16.000         28         0
     Vset[6]     1.000      2.643     7.000         28         0
     Vset[7]     2.000      7.393    14.000         28         0
     Vset[8]      0.00      10.33     50.00         28         1      Skew
     Vset[9]      9.00      17.64     50.00         28         0      Skew
    Vset[10]     20.00      64.82    176.00         28         0
    Vset[11]     14.00      39.57     77.00         28         0
    Vset[12]      0.00      34.00     86.00         28         5
    Vset[13]     3.000      5.679    12.000         28         0
    Vset[14]     26.00      57.64    128.00         28         0
```

Since we have a units-by-variates set of data, rather than a similarity matrix, we need to get the similarities among the 28 brooches. We can look at the details of that operation a little later; for the moment we shall use a simple method. The FSIMILARITY statement given below will form the similarity matrix Similars, using all 14 variates as its input; however, it must know how each variate is to be treated. The TEST parameter contains this information in a coded form: the value 5, which will be recycled 14 times for each of the variates, indicates that each variate is to be treated in the same way. Exactly what code 5 means will be explained in Section 7.4. The FSIMILARITY statement prints some summary information about each of the variates, and the overall size of the similarity matrix. An FSIMILARITY statement can also print the similarity matrix, if the PRINT option is set; two different formats are provided, of which the very compact form has been chosen here. This prints the first digit after the decimal point for each value, and does not separate the digits at all; self-similarity values are indicated by minus signs down the diagonal. This form is especially useful when there is a large number of units, for example up to about 100.

```
32  FSIMILARITY [PRINT=similarity; STYLE=abbreviated; SIMILARITY=Similars] \
33     Vset[]; TEST=5

Variate        Minimum      Range
     Vset[1]     10.00      84.00
     Vset[2]     7.000     19.000
     Vset[3]     1.000      6.000
     Vset[4]     6.000      4.000
     Vset[5]     5.000     11.000
     Vset[6]     1.000      6.000
     Vset[7]     2.000     12.000
     Vset[8]        0.        50.
     Vset[9]     9.000     41.000
    Vset[10]     20.00     156.00
    Vset[11]     14.00      63.00
    Vset[12]        0.        86.
    Vset[13]     3.000      9.000
    Vset[14]     26.00     102.00
```

```
Similarity matrix computed, length = 406

** Abbreviated similarity matrix **

  1   -
  2   5-
  3   78-
  4   668-
  5   7799-
  6   69988-
  7   679898-
  8   6897898-
  9   57999888-
 10   879998888-
 11   8899998989-
 12   68989989999-
 13   699898889899-
 14   5798988898899-
 15   68989898988999-
 16   689999999899999-
 17   6898998898999999-
 18   78989999989989999-
 19   789898988899899999-
 20   9677676768876666767-
 21   68989999989999999996-
 22   589798999789899989959-
 23   6899988899999999999798-
 24   56999887988889998985988-
 25   779898989999999999979998-
 26   6898999898998999999699999-
 27   58978998878888998995998889-
 28   777778677788666777787666777-
```

7.3 Average linkage cluster analysis

We can now move on to the analysis, which is done by the HCLUSTER directive. One option of HCLUSTER is the choice of method; here average linkage clustering has been chosen. Now we simply specify the identifier of the similarity matrix:

```
HCLUSTER [PRINT = amalgamations,dendrogram; \
    METHOD = average] Similars
```

The output for average linkage clustering is in three parts; the first being the information about the merging of clusters.

```
**** Average linkage cluster analysis ****

** Merging clusters **

    18     21   98.5
    14     15   98.4
     3     16   98.2
     7     19   98.2
     3     26   97.7
    13     23   97.3
    18     22   97.2
     5      9   96.9
    12     25   96.7
     5     14   96.6
     3      5   95.8
    13     17   95.6
    10     11   95.0
     3     12   95.0
     7     27   94.8
     3     18   94.2
     1     20   92.9
     3      8   92.3
     4     24   91.9
     2      6   91.0
    10     13   91.0
     3      7   89.4
     4     10   87.6
     2      3   86.8
     2      4   83.3
     1     28   79.9
     1      2   71.4
```

Initially all the brooches are assumed to be in separate groups. Now the analysis starts by examining the similarity matrix and determining that the largest similarity value (0.985) is between the 18th and 21st brooches. These are merged into a single group, then the similarities between this group and all the other brooches are calculated. Because we have chosen the average linkage method, the similarity between the *i*th brooch and the new group is the average of (a) the similarity between the *i*th brooch and the 18th brooch, and (b) the similarity between the *i*th brooch and the 21st brooch. If a different method were being used, the calculation of the new similarities would be different. Now the analysis finds the largest similarity among the 27 remaining groups, which is between the 14th and 15th brooches; these are merged, and new similarity values are calculated. The similarity between the two groups (18, 21) and (14, 15) is the average of the similarities (18, 21) with 14, and (18, 21) with 15. The next join is between the third and 16th brooches, and it can be seen that the third brooch is also involved in the fifth merge, with the 26th brooch. Actually, the fifth merge is between the group containing the third brooch, (3, 16), and the 26th brooch. In the merging-clusters output a group is always represented by the smallest unit number in the group, which is always the value in the left-most column of the output. The right-most column of the output gives the similarity at which each merge occurred. These are in descending order because the method is always looking for the largest similarity to determine the next merge.

Although the output above is complete, it is hard to see what the different groups are at any stage. The next part of the output from the HCLUSTER statement provides a summary of the grouping at different levels of similarity. HCLUSTER does not print any details of clusters until some clustering has occurred, and stops printing once a single group has been obtained. The similarity levels are presented as percentages; by default the summaries are provided at equal intervals chosen to provide a reasonable number of summaries of the grouping. Looking through the output given below, it would seem that the 28th brooch is (relatively) quite different from the others; also the first and 20th brooches are different from the others, although fairly similar to each other. Whether the remaining 25 brooches form a single group (for example at level 80%), or are in two groups (85%), or more than two groups, is uncertain. If two or three groups had formed at a high similarity and not merged until a much lower similarity we could be more sure.

```
**** Hierarchical clusters ****

  ** Level      95.0

   3    16    26     5     9    14    15

  12    25

  18    21    22

   7    19

  13    23    17
** Ungrouped
   1    20    28     2     6     8    27     4    24    10
  11

  ** Level      90.0

   1    20

   2     6

   3    16    26     5     9    14    15    12    25    18
  21    22     8

   7    19    27

   4    24

  10    11    13    23    17
** Ungrouped
  28

  ** Level      85.0

   1    20

   2     6     3    16    26     5     9    14    15    12
  25    18    21    22     8     7    19    27

   4    24    10    11    13    23    17
** Ungrouped
  28
```

```
  ** Level      80.0

   1   20

   2    6    3   16   26    5    9   14   15   12
  25   18   21   22    8    7   19   27    4   24
  10   11   13   23   17
** Ungrouped
  28


** Level      75.0

 1   20   28

 2    6    3   16   26    5    9   14   15   12
25   18   21   22    8    7   19   27    4   24
10   11   13   23   17

** Level      70.0

 1   20   28    2    6    3   16   26    5    9
14   15   12   25   18   21   22    8    7   19
27    4   24   10   11   13   23   17
```

The final part of the output from HCLUSTER is a *dendrogram*. This combines aspects of both displays shown above and, probably for this reason, is extremely popular. However, it is very easy to misinterpret dendrograms, as is discussed by Digby (1986), so it is advisable to look at all three components of the output from HCLUSTER. The dendrogram for this analysis is printed below.

```
**** Dendrogram ****
** Levels    100.0  90.0  80.0  70.0

           1  .....
          20  .....)........
          28  ..............)..
           2  .....            )
           6  .....)..         )
           3  ..      )        )
          16  ..)     )        )
          26  ..)     )        )
           5  ..)     )        )
           9  ..)     )        )
          14  ..)     )        )
          15  ..)     )        )
          12  ..)     )        )
          25  ..)..   )        )
          18  ..   )  )        )
          21  ..)  )  )        )
          22  ..)..)  )        )
           8  .....)..)        )
           7  ..      )        )
          19  ..)..   )        )
          27  .....)..)..      )
           4  .....      )     )
          24  .....)..   )     )
          10  ..      )  )     )
          11  ..)..   )  )     )
          13  ..   )  )  )     )
          23  ..)  )  )  )     )
          17  ..)..)..)..).....)...........
```

Down the left-hand side of the dendrogram are the unit numbers of the brooches. Across the top of the display are the different similarity levels at which merging takes place; again the similarities are expressed as percentages. The merging of groups is indicated by the brackets in the body of the dendrogram. For example, from the top few lines of the display, we see that units 1 and 20 merge somewhere between 100% and 90% similarity; this group is joined by the 28th brooch somewhere after 80% similarity. The long line of brackets down the right of the display indicates that this small group, of three brooches, joins the remaining 25 brooches, which have already been joined into one group. Whether this suggests that the group (1, 20, 28) is markedly different from the other 25 brooches is doubtful, but should be borne in mind in the analyses that follow.

7.4 Ways of forming similarity matrices

Doran and Hodson (1975) give scale drawings of the brooches; three of them are larger than the others, and this is the group (1, 20, 28) that we have identified already. So it would seem that a large component of the similarity value is dependent on the size of the brooch. If we are interested in differences in style, rather than size, we need to form a similarity matrix that does not depend on the overall size. To do this requires some knowledge of what the FSIMILARITY directive does.

Gower (1971) gave some general guidelines on forming similarity matrices; in particular, he suggested that the overall similarity should be the average of similarities formed from each of the p variables separately. If s_{ij} is the similarity of the ith and jth units, then

$$s_{ij} = \frac{1}{p} \sum_k s_{ij,\,k}$$

where $s_{ij,k}$ is the similarity of the ith and jth units for the kth variable. This is convenient because it allows different types of variable to be handled in different ways. It also provides a neat way of dealing with missing values, by ignoring any values of $s_{ij,k}$ that involve missing values.

In Genstat there are five different ways in which $s_{ij,k}$ can be defined. If x_{ik} is the ith data value for the kth variable, which has range r_k, these methods are given below.

1. $s_{ij,\,k} = 1$ if $x_{ik} = x_{jk}$ and $x_{ik} \neq 0$
   $= 0$ otherwise

2. $s_{ij,\,k} = 1$ if $x_{ik} = x_{jk}$
   $= 0$ otherwise

3. $s_{ij,k} = 1 - \dfrac{|x_{ik} - x_{jk}|}{r_k}$

4. $s_{ij,k} = 0 \text{ if } x_{ik} = x_{jk} = 0$

$$= 1 - \frac{|x_{ik} - x_{jk}|}{r_k} \text{ otherwise}$$

5. $s_{ij,k} = 1 - \left(\dfrac{x_{ik} - x_{jk}}{r_k}\right)^2$

The first and second of these are for qualitative information, where one can say that two things are similar only if they have the same value. An example of this sort of measurement would be a variable indicating colours. The latter three are for quantitative variables; they define similarity relative to the maximum possible difference, the range. The fifth form is on a quadratic scale; it is simply one minus the squared Pythagorean distance (see Chapter 6, Section 6.3) measured on variables normalized to have a range of one. The third is related to the Manhattan (or city-block) metric. The first and fourth methods are provided as variants of the second and third because it is sometimes advisable to ignore matches of zero with zero. This occurs, for example, with ecological data where the mutual absence of a species from two sites does not imply the similarity of those sites.

In trying to construct a similarity matrix that is independent of brooch size it is important to recognize that the 14 variates are of three types. One variate is the total length, and nine others are also length measurements. If these nine are divided by the total length we should achieve some independence of size. With continuous data such as these length ratios the Pythagorean-related method (type 5 above) is commonly used. Variable 13 contains the number of coils on the brooch; looking through these it can be seen that most of the values are 4 or 6, but that three brooches have 10 or 12 coils. It seems reasonable to use this variable dichotomously; we shall say that two brooches are similar for this variable if they both have less than eight coils, or both have more than eight. Variables 3, 4, and 6 are angles, given in 10-degree intervals; the linear-scale qualitative method (type 3) seems a reasonable choice. Lastly, we shall consider the relative weights of the variables. Since there are nine length ratios, most of the information will come from these. However, three times as much information will come from the angular measurements as from the coil information if the variables are given equal weight. To increase the contribution of the latter we shall include it three times. The Genstat statements to form the new similarity matrix are as follows.

```
35  CALCULATE Vset[1,2,5,7,8...12] = Vset[1,2,5,7,8...12] / Vset[14]
36  &   Vset[13] = Vset[13] > 8
37  FSIMILARITY [SIMILARITY=Similars] Vset[1...13,13,13]; \
38     TEST=2(5,3),5,3,(5)9

Variate       Minimum      Range
     Vset[1]   0.2667     0.5491
     Vset[2]   0.1172     0.3717
     Vset[3]    1.000      6.000
     Vset[4]    6.000      4.000
     Vset[5]  0.08182    0.16818
     Vset[6]    1.000      6.000
     Vset[7]  0.02727    0.28852
     Vset[8]       0.         0.
     Vset[9]   0.1636     0.2486
    Vset[10]   0.3070     2.9522
    Vset[11]   0.2593     1.4330
    Vset[12]       0.         1.
    Vset[13]       0.         1.
    Vset[13]       0.         1.
    Vset[13]       0.         1.

Similarity matrix computed, length = 406
```

7.5 Other methods of clustering

Genstat provides five methods of cluster analysis. These differ in the way in which the similarities between a new group and all the existing groups are derived. For the analysis below we shall use the default method, which is *single linkage cluster analysis*. The permutation of the units in the dendrogram can be stored in a variate, as here. Single linkage analysis produces different output from the other methods. The first part of the output is the *minimum spanning tree*.

```
 39  HCLUSTER [PRINT=amalgamations,dendrogram] Similars; PERMUTATION=Clord
 **** Single linkage cluster analysis ****

 **** Minimum spanning tree ****
Similars

       1...... 20...... 11......  3...... 10...... 12
          93.9      92.5  (  93.2  (  97.3  (  97.6
                          (        (        (
                          (        (        (...... 13......  2
                          (        (        (  78.4  (  85.6
                          (        (        (        (
                          (        (        (        (...... 23
                          (        (        (           95.5
                          (        (        (
                          (        (        (...... 17
                          (        (        (  94.7
                          (        (        (
                          (        (        (...... 25
                          (        (           95.7
                          (        (
                          (        (...... 16......  4
                          (           96.0  (  93.0
                          (                 (
                          (                 (...... 26......  5...... 21......>
                          (                    96.0      97.0      97.8
                          (
                          (......  6...... 28
                          (  92.9     90.6
                          (
                          (......  8
                             93.6
```

```
       21...... 15...... 14......   9
        (  96.3  (  96.1     94.2
        (        (
        (        (...... 24
        (           94.5
        (
        (...... 18...... 27...... 19......   7
        (  95.9     94.2     95.8     95.0
        (
        (...... 22
           97.1
** Total length      2540.8
```

A spanning tree is a series of connections among the units with two properties: first, it must be possible to move along these connections from every unit to every other unit (thus it "spans" the units); second, there must be no closed loops among the connections (so that it conforms to the mathematical idea of a tree). The minimum spanning tree (MST) is the spanning tree with the shortest possible total length of all the connections. The length of a connection is the distance, that is one minus the similarity, between the two units that it joins.

For example, in the MST printed above there are connections between the first and 20th brooches, and between the 20th and the 11th. The 11th brooch is also connected to the sixth and eighth brooches. The similarity associated with each link is printed underneath it; the total of these values is 2540.8, which is printed below the tree. Note that the output page was too narrow to print the tree completely, so that a part of it, the links from the 21st brooch onwards, is printed separately.

Single linkage clustering differs from the average linkage method used earlier in the definition of the similarity of a new group, formed by merging two existing groups, with all the other groups. This is taken as the maximum of the two similarities involved. For example, the first merge is of the fifth and 21st brooches; now the similarity of this group with any other brooch, say the 26th, is taken as the maximum of the similarities between the fifth and 26th or between the 21st and 26th. (In fact it is the former of these, and is 97%). Because of this method of specifying new similarities, most of the clustering occurs at quite high similarity values, as shown below.

```
**** Hierarchical clusters ****

  ** Level      95.0

   3    10    12    16    26     5    21    22    15    14
  18    25

  27    19     7

  13    23
** Ungrouped
   1    20    11     8    17    24     9     4     6    28
   2

  ** Level      90.0

   1    20    11     8     3    10    12    16    26     5
  21    22    15    14    18    25    17    24     9    27
  19     7     4     6    28

  13    23
** Ungrouped
   2
```

```
** Level      85.0

  1    20    11     8     3    10    12    16    26     5
 21    22    15    14    18    25    17    24     9    27
 19     7     4     6    28

 13    23     2


** Level      80.0

  1    20    11     8     3    10    12    16    26     5
 21    22    15    14    18    25    17    24     9    27
 19     7     4     6    28

 13    23     2


** Level      75.0

  1    20    11     8     3    10    12    16    26     5
 21    22    15    14    18    25    17    24     9    27
 19     7     4     6    28    13    23     2
```

Note that quite a large group, of ten brooches, has been formed at a similarity as high as 96%. Note also that the final join is of the three brooches with the large number of coils to the remaining brooches.

The dendrogram from single linkage cluster analysis is printed below. Again the distinct group of three brooches (13, 23, 2) can be seen; there also appears to be a separate group of four brooches (1, 20, 11, 8) at the top of the dendrogram. These groupings are different from the grouping of large brooches found earlier, suggesting that the data transformations have been effective in removing the size component.

```
**** Dendrogram ****
** Levels    100.0  90.0  80.0

                1  .....
               20  .....)
               11  .....)
                8  .....)
                3  ..   )
               10  ..)  )
               12  ..)  )
               16  ..)  )
               26  ..)  )
                5  ..)  )
               21  ..)  )
               22  ..)  )
               15  ..)  )
               14  ..)  )
               18  ..)  )
               25  ..)..)
               17  .....)
               24  .....)
                9  .....)
               27  ..   )
               19  ..)  )
                7  ..)..)
                4  .....)
                6  .....)
               28  .....)........
               13  ..           )
               23  ..).....     )
                2  ........).....)........
```

7.6 Printing re-ordered similarity matrices

In the HCLUSTER statement above the order of the brooches used for the dendrogram was saved in the variate Clord. This can now be used in the FSIMILARITY directive to print the similarity matrix with its rows and columns reordered to match the dendrogram.

```
FSIMILARITY [PRINT = similarity; STYLE = abbreviated; \
    SIMILARITY = Similars; PERMUTATION = Clord]
```

Because there is no list of variates given after the options the directive is not being used to form a similarity matrix; it will simply print the matrix Similars.

```
 40       FSIMILARITY [PRINT=similarity; STYLE=abbreviated; \
 41          SIMILARITY=Similars; PERMUTATION=Clord]

** Abbreviated similarity matrix **

  1   -
 20   9-
 11   89-
  8   889-
  3   8898-
 10   88989-
 12   889999-
 16   7788999-
 26   77889889-
  5   778898899-
 21   8788998999-
 22   77898888899-
 15   778898899999-
 14   7687898989989-
 18   87889989999998-
 25   878899998899999-
 17   8788999888888989-
 24   65778778888899887-
  9   668888998888998988-
 27   8788888888888899888-
 19   88889888889898998889-
  7   767788788888889878899-
  4   7788888999889888888887-
  6   88989889988887888778879-
 28   788877777776667775677579-
 13   6576777766766777756666665-
 23   55667777777677777666767649-
  2   677666666565555666456656 7788-
```

This output is more useful than a printed copy of the matrix without the reordering because it allows internal consistencies of groups to be checked.

7.7 Using principal coordinate analysis with classification

When principal coordinate analysis (PCO) was introduced in Chapter 6, Section 6.2 it was mentioned that the symmetric matrix used as input to PCO should contain similarities, rather than distances. We have used similarity matrices throughout this chapter for cluster analysis; they can also be used as input to PCO. This will represent the brooches such that the squared distance between the points for two brooches is proportional to one minus their similarity. Continuing the previous Genstat program to do this gives the following output.

```
  42  LRV [ROWS=28; COLUMNS=3] L
  43  PCO [PRINT=roots,scores; NROOTS=2] Similars; L

43..................................................................

*****  Principal coordinates analysis  *****

***  Latent Roots  ***

                     1           2
                 1.410       1.122

***  Percentage variation  ***

                     1           2
                 28.64       22.78

***  Trace  ***

     4.923

*  Some roots are negative - non-Euclidean distance matrix  *


***  Latent vectors (coordinates)  ***

                       1           2
          1       0.2231     -0.2468
          2       0.6041      0.2526
          3      -0.0063     -0.0648
          4      -0.0724     -0.0317
          5      -0.1766     -0.0072
          6       0.1392     -0.1997
          7      -0.2392      0.0264
          8       0.1127     -0.1450
          9      -0.2261      0.0902
         10       0.0174      0.0004
         11       0.1720     -0.1744
         12       0.0348     -0.0220
         13       0.3248      0.5258
         14      -0.2392      0.1250
         15      -0.2505      0.0575
         16      -0.0995     -0.0212
         17       0.0021      0.0211
         18      -0.1273     -0.0341
         19      -0.0849     -0.0418
         20       0.3486     -0.3293
         21      -0.1566     -0.0292
         22      -0.1709     -0.0155
         23       0.2488      0.5618
         24      -0.3876      0.1384
         25      -0.0726      0.0256
         26      -0.1191     -0.0477
         27      -0.1246     -0.0363
         28       0.3259     -0.3784

  44  VARIATE [NVALUES=28] Sc[1...3]
  45  CALCULATE Sc[] = L[1]$[*; 1...3]
  46  FACTOR [LEVELS=28; VALUES=1...28] Fg
  47  GRAPH [EQUAL=scale; NROWS=37; NCOLUMNS=61] Sc[2]; Sc[1]; SYMBOLS=Fg
```

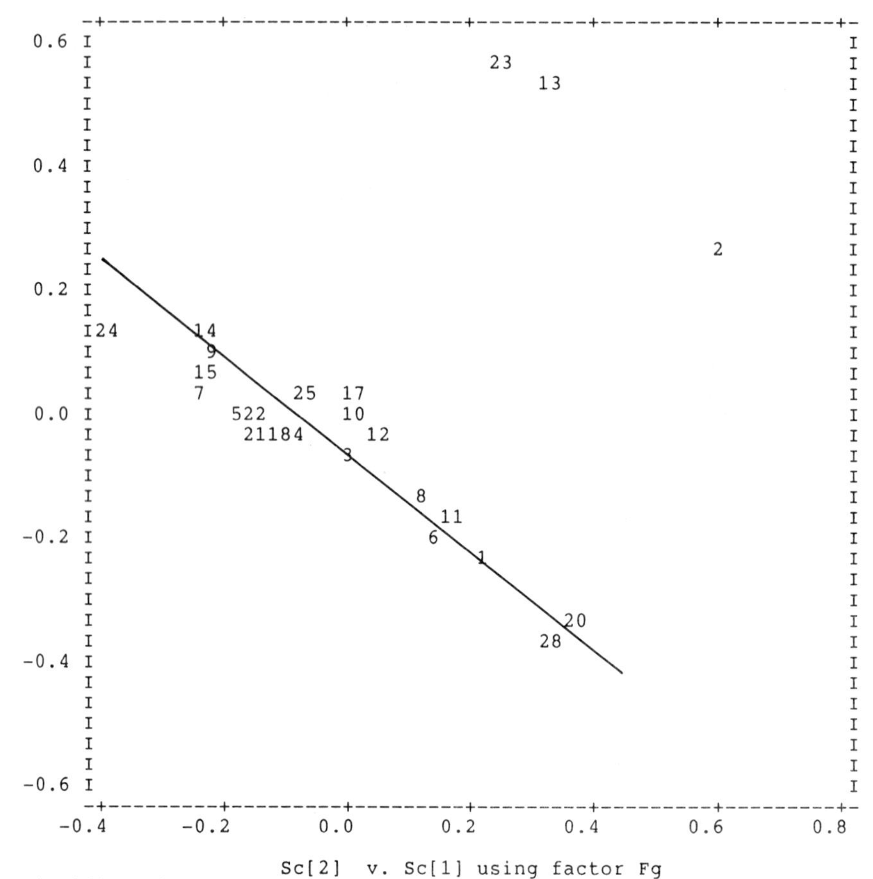


```
Points coinciding with 4
16 19

Points coinciding with 18
26 27
```

In accord with the dendrogram of Section 7.5 the separation of the 23rd, 13th, and second brooches from the others is quite striking. Also the group (1, 11, 20, 28) occurring at the top of the dendrogram can be seen to the bottom-right of the main block of points. To investigate this further requires the use of canonical variate analysis, described in Chapter 5, Section 5.6, which might indicate the variables that are contributing to differences between these two groups and the remaining 21 brooches.

Diagrams from PCO can be useful in a number of ways to complement a classification. For example, closed curves can be drawn around identified groups; this can be done at various similarity values to represent the hierarchy of grouping.

Often it is useful to draw the minimum spanning tree on a graph of PCO coordinates. On the graph shown above this would result in a mess; however, we can obtain a different graph and use that. One part of the MST consists of the small group of three brooches (2, 13, 23) and is connected to the rest of the tree by a link with similarity 78.4. If we ignore that part of the tree we need points only for the 25 brooches that are roughly on a line at $-45°$, which we have drawn on the PCO graph. We can calculate one new coordinate from the two old coordinates by moving the origin to the centroid of the 25 points and then turning the axes through 45°. This new coordinate can be plotted with the third coordinate from PCO and the MST for the subset of 25 brooches drawn in by hand. To centre and rotate the points we first use RESTRICT to exclude the group of three brooches. This is done using the .NI. operator to indicate that only those units for which the factor Fg has a value "not in" the set of values (2, 13, 23) should be used. Second, the means of the variates of scores are subtracted. The rotation is simply Sc[1] $\times$ cos(45) $-$ Sc[2] $\times$ sin(45); although it is important to remember that Genstat uses angles expressed in radians rather than degrees, so that 45° is given as $\pi/4$. The Genstat output, including the statements needed to get the new coordinate, is shown below.

```
48  RESTRICT Sc[],Fg; Fg .NI. !(2,13,23)
49  CALCULATE Sc[1,2] = Sc[1,2] - MEAN(Sc[1,2])
50  & Sc12 = Sc[1]*COS(3.14159/4) - Sc[2]*SIN(3.14159/4)
51  GRAPH [EQUAL=scale; NROWS=37; NCOLUMNS=61] Sc[3]; Sc12; SYMBOLS=Fg
```

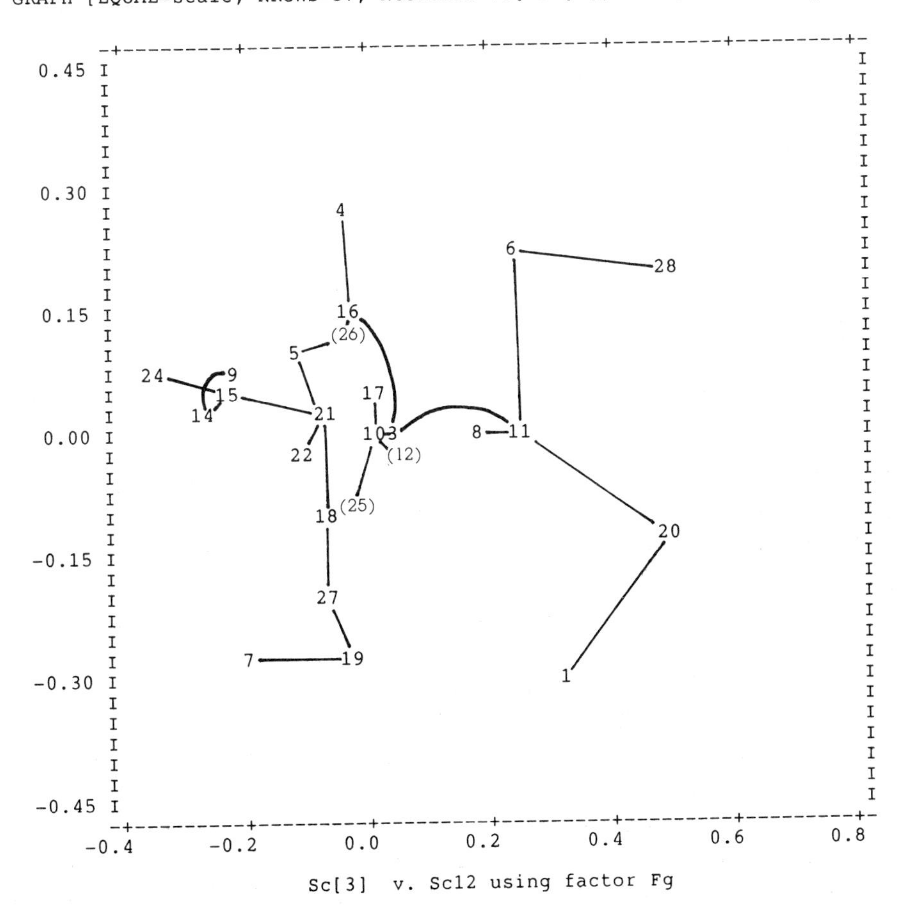

```
Points coinciding with 3
12

Points coinciding with 18
25

Points coinciding with 16
26
```

The minimum spanning tree can also be used to show the progression of single linkage clustering. The analysis actually finds the links of the MST, in decreasing order of similarity, and uses these to define the merging of groups. At any stage in the analysis, a set of points that are joined, either directly or indirectly, is a single group. For example, the link with smallest similarity in the MST is that between the 10th and 13th brooches. By the time that link is made all the links among the 25 brooches shown above have been made; also the two links needed to form the group (2, 13, 23). The last link is the one that merges the two groups into one. This dualism between the MST and single linkage clustering can be used to good effect by preparing a number of graphs, from PCO, that show different stages of the clustering.

7.8 Exercises

7(1) Using the data from Chapter 5, Exercise 5(1), write a program to construct a similarity matrix for the 26 insects, and then to cluster them using any method other than single linkage cluster analysis. You should consider whether any preliminary transformation of the data might be beneficial, and also which type(s) of similarity might be appropriate.

7(2) Using the similarity matrix formed in Exercise 7(1), write a program to use single linkage cluster analysis. Prepare several copies of a graph of the scores for the 26 insects, from either principal components analysis or principal coordinate analysis, and draw partially complete minimum spanning trees on these to show several stages of the clustering.

7(3) The GTHRESHOLD and GROUPS options of the HCLUSTER directive can be used in combination to save a clustering into some number of groups, using the cluster analysis of either Exercise 7(1) or Exercise 7(2), write a program to save a clustering into three groups and compare this with the *a priori* grouping given in Chapter 5, Exercise 5(3), preferably using the TABULATE directive.

8 Programming in the Genstat language

This chapter describes some of the features of Genstat that are useful for general programming. Because the Genstat language provides directives for program construction, as well as powerful facilities for calculations and data manipulation, it can be used for general-purpose programming. Thus Genstat can be used for almost any type of statistical analysis, irrespective of whether the analysis is provided directly within Genstat. An introduction to some of the programming aspects of Genstat, in particular FOR-ENDFOR loops and IF-ELSE-ENDIF constructions, is given in Chapter 8 of *Genstat 5: an Introduction*. This chapter also describes a few of the less commonly used features of some Genstat directives, such as PRINT and CALCULATE.

8.1 Structured programming

An important aspect of any computer language is the manner in which it allows the user to control the order in which statements are executed: there are two general ways in which languages provide the ability to alter this flow of execution. The first of these uses special statements to label points in the program, and statements that allow the user to continue execution from a labelled point, as in Fortran IV. The more modern style is to use the concepts of *structured programming*, as in the computer language Pascal. It is generally considered that programs written using structured programming are clearer to the reader and more likely to be correct in the logic of their construction.

Genstat has several directives that are used for structured programming, and does not have the facilities for the "labels and goto" approach. The main constructions of structured programming can be subdivided into two types: those that allow repetition and those that allow conditional execution.

In Genstat the FOR and ENDFOR directives are used for specifying repetition. This can be of the type where the number of repetitions is known in advance, as for example in the first program in Chapter 1. Alternatively, these directives can be used when the number of repetitions required to satisfy some condition is not known, as illustrated in this chapter.

Conditional execution is provided in Genstat by the IF, ELSIF, ELSE, and ENDIF directives which allow the selection of a course of action depending on one or more conditions. A special case occurs when the selection is simply to choose the first course of action, or the second course of action, and so on, from a list of possibilities. Although this could be specified using IF and ELSIF statements, it is easier, and clearer, to use the CASE, OR, ELSE, and ENDCASE directives in Genstat. The statements following the ELSE statement, when it is used inside a CASE-ENDCASE structure, provide a "catch-all" to use if the selecting expression in the CASE statement does not correspond to an existing set of statements. For example, the following sets of statements are equivalent:

```
IF Selector == 1              CASE Selector
  ...                           ...
ELSIF Selector == 2           OR
  ...                           ...
ELSIF Selector == 3           OR
  ...                           ...
ELSE                          ELSE
  ...                           ...
ENDIF                         ENDCASE
```

Even when using a structured-programming style, it is sometimes useful to use a special form of the "labels and goto" approach. This is when some condition has become true and it is required to exit from the current control structure; that is, to go to its end and continue execution from there. Genstat has the EXIT directive to cater for this, as will be described in Section 8.4. For example, when using a FOR-ENDFOR loop to analyse several variables, it might be convenient to exit from one pass of the loop if an error is detected for one of the variables, and then to continue going through the loop for the other variables.

8.2 Designing a program

Rings of standing stones occur at many places around Britain. Several of them appear to have the stones positioned around the circumference of a circle, or several circles as at Stonehenge; however, some of the rings appear to be flattened circles, or ellipses. In order to examine how well any particular ring can be fitted by a circle, we shall first need to obtain the geographical coordinates of the stones. Then the problem is to fit a circle to the coordinates. Finally, the fitted circle, and the stone positions, will need to be displayed.

A diagram of a ring of 12 stones at the House of Aquahorthies, near Inverurie in the Highlands of Scotland, is given in Figure 12.3 of *Megalithic Sites in Britain* (Thom 1979). The site has also been surveyed, and the coordinates of the stones have been digitized and supplied to us by C. Ruggles and S. Kruse (University of Leicester); the digitized coordinates are listed below.

1.5237236E+00	1.0512443E+00
3.1040440E+00	−4.0295219E−01
4.0263481E+00	−2.2693760E+00
3.6383057E+00	−4.2206721E+00
2.4089622E+00	−5.7882729E+00
1.5897942E−01	−6.1882696E+00
−9.2122650E−01	−5.9324398E+00
−1.9452114E+00	−5.7220306E+00
−3.7378788E+00	−3.6405981E+00
−3.6993866E+00	−1.4707320E+00
−2.6538162E+00	2.5678873E−01
−7.4598503E−01	1.1301274E+00

Fitting a circle to the positions of the stones involves finding estimates of the coordinates of the centre and of the best-fitting radius. There is no directive in Genstat for this, but it is not very difficult to derive least-squares equations for the best estimates of the coordinates of the centre and of the radius; these can then be used to form part of a Genstat program. Unfortunately, the equations for the coordinates of the centre assume that the best radius is known, and the equation for the radius assumes that the coordinates of the centre are known. The only way to solve the least-squares equations is to make an initial guess at the coordinates of the centre (or at the radius), and then repeatedly alternate between estimating the radius and then the centre (or vice versa) until the repetitions converge to a single solution. Each of these repetitions is termed an iteration.

It is fairly easy to get an initial guess at the centre of the circle by averaging the x- and y-coordinates, so this alternative has been chosen. There are various ways in which we could check for convergence of the iterative process. The method used here calculates a residual sum of squares at each iteration and stops iterating when the change in this, from one iteration to the next, becomes relatively small, so that the process converges. It may be useful if the iterative process reports its progress, so we shall allow for such reporting, if required. It is possible for iterative processes never to converge: we shall need to guard against this, although it is unlikely to happen for this particular method.

Once the iterative process has finished, we shall need to display the results. Various possibilities exist for this; for example, drawing a line-printer graph or using the high-resolution graphical output from Genstat. There is always the possibility, mentioned above, that the process has not converged to a solution and we shall need to be able to report this as well, and then to suppress any graphical output.

The complete scheme is shown in the flow diagram of Figure 8.1.

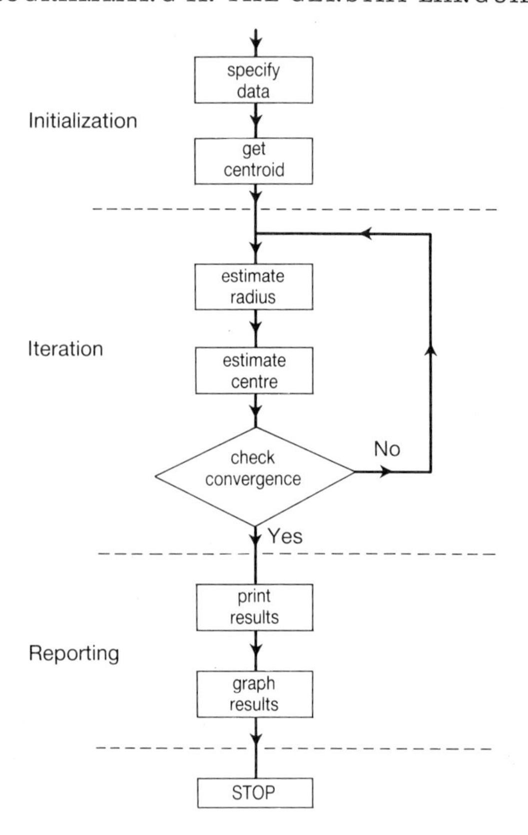

Figure 8.1. Flow diagram of the iterative fitting process.

You will notice that we have looked at the complete problem and split it up into three sub-problems: getting the data, fitting the circle, and reporting the results. The problem of fitting the circle has itself been split into smaller problems: getting an initial estimate of the centre, and iterating until the estimates of the centre and radius have converged. This approach is sometimes called *top-down programming*, because it starts at the top and defines the single problem in terms of several smaller problems, each of which may then also be split up. Further problem subdivision can occur, until one gets to the stage of writing actual statements of the chosen computer language. For the circle-fitting problem, and using Genstat to solve it, it is fairly easy to move to the stage of writing a program from the present analysis of the problem.

Berman and Culpin (1986) describe several approaches to the fitting of circles to a set of points. In particular, they point out that least-squares estimation, as we are using in this chapter, can be unreliable and may converge to an absurd solution. They suggest that a more reliable method is least-squares-squared estimation which has a unique algrebraic solution given by:

$$\hat{a} = \bar{x} + \frac{\Sigma v_i^2 \Sigma u_i (u_i^2 + v_i^2) - \Sigma u_i v_i \Sigma v_i (u_i^2 + v_i^2)}{2 \{\Sigma u_i^2 \Sigma v_i^2 - (\Sigma u_i v_i)^2\}}$$

$$\hat{b} = \bar{y} + \frac{\Sigma u_i^2 \Sigma v_i (u_i^2 + v_i^2) - \Sigma u_i v_i \Sigma u_i (u_i^2 + v_i^2)}{2 \{\Sigma u_i^2 \Sigma v_i^2 - (\Sigma u_i v_i)^2\}}$$

$$\hat{r} = (\bar{x} - \hat{a})^2 + (\bar{y} - \hat{b})^2 + \frac{1}{n} \Sigma (u_i^2 + v_i^2)$$

where $u_i = x_i - \bar{x}$ and $v_i = y_i - \bar{y}$.

8.3 Initialization: getting the data

The first part of the program defines the structures needed for the program, and reads in the data: it is shown below.

```
  1    JOB 'Fitting circles'
  2       " This Job takes the (x,y) coordinates of a set of points and
 -3         finds the least-squares estimates of the parameters of the
 -4         best-fitting circle to those points.  The parameters are the
 -5         centre of the circle (a_hat,b_hat) and the radius (r_hat).    "
  6    SCALAR Lots,Thresh,Report,Graphopt; VALUE=100,0.0001,1,2
  7    OPEN 'Circle.dat','Circle.lis'; CHANNEL=2; FILETYPE=input,output
  8    READ [CHANNEL=2; SETNVALUES=yes] X,Y

  Identifier   Minimum      Mean   Maximum    Values   Missing
           X  -3.73788   0.09640   4.02635        12         0
           Y    -6.188    -2.766     1.130        12         0
  9    CLOSE 2; FILETYPE=input
 10    POINTER [VALUES=A_hat,B_hat] Centre
 11    SCALAR Centre[],R_hat,Rss,Oldrss,Change,Iterno; VALUE=0
 12    VARIATE [NVALUES=X] Xadj,Yadj,R
 13    CALCULATE Centre[] = MEAN(X,Y)
 14    CALCULATE Oldrss = SUM(X**2 + Y**2)
 15    IF Report
 16       PRINT [CHANNEL=2; IPRINT=*; ORIENTATION=across] \
 17          !T(iteration,A_hat,B_hat,R_hat,Residual,Criterion); \
 18          FIELDWIDTH=12
 19       & [SQUASH=yes] Iterno,Centre[]; \
 20          FIELDWIDTH=12; DECIMALS=0,(*)2
 21    ENDIF
```

After the descriptive comment, Lines 6 and 7 set up various aspects of the job that may need to be changed from one run to another. Four scalars, declared on Line 6, are used to control the action of the program: Lots specifies the maximum number of iterations; Thresh holds the value used to determine when the iterative process has converged; Report specifies whether a report of the iterative process is to be printed to a secondary output file; Graphopt specifies what type of graphical display, if any, is required. Although it is not necessary to declare scalars to hold these constants, it is sensible to do so because their specification can be kept in one place in the program, rather than being scattered around. This makes it easy to change the action of the program without having to look through all of it to find the places where changes are needed. For general programs that are to be used by several people it is probably wise to use extra comments to explain any changes that may need to be made.

The OPEN statement on Line 7 opens the secondary input file that contains the data, and a secondary output file to contain the results. You should note that the second parameter list, which specifies the channel number, is the single value 2. There is no cause for confusion here because the various channel numbers used for input are completely distinct from those used for output, and from any others, for example those used in Chapter 9 for procedure libraries.

The data are input on Line 8; here the SETNVALUES option has been set to 'yes' so that Genstat will decide the length of the input structures from the amount of data present.

Within the program the estimate of the centre of the circle will be held using two scalars, A_hat and B_hat, for its coordinates. They are referred to at several points in the program, and always together. Rather than spell out the list A_hat,B_hat at each point it is more convenient to use them as the elements of the pointer Centre, declared on Line 10. An additional advantage in using a pointer is the reduction in the chance of a typing mistake going unnoticed: whereas Genstat might not object to

```
CALCULATE A_aht,B_hat = MEAN(X,Y)
```

it almost certainly will to

```
CALCULATE Cetnre[] = MEAN(X,Y)
```

as it will not know how many structures are represented by the newly introduced pointer Cetnre.

Although it is often unnecessary to declare structures in Genstat explicitly, it can be useful to do so, especially if the program may be used by other people. Various scalars are needed in the program and these are declared on Line 11. To represent the scalars A_hat and B_hat, the reference Centre[] is used: this stands for all the structures to which Centre points, and is a convenient shorthand for Centre[1,2] which would mean the same thing here. Although only the scalar Iterno, to hold the iteration count, needs initializing, all the scalars will be given the value 0. Three variates, with the same length as the input data, are needed during the iterative process and these are declared on Line 12. Here we have not needed to calculate the length of the input variates to use in the declaration; the length of the three variates will be taken from the variate X.

To start the iterative process we need an initial estimate of the centre of the circle. This is the centroid position; it is calculated on Line 13 and stored in the scalars A_hat and B_hat. In order for the convergence check to work on the first iteration, we need to supply some relatively large value for the old residual sum of squares. A suitable value is calculated on Line 14.

Finally, before the calculation of the parameters of the circle, the column headings for the reporting need to be printed, provided that the program is to report the progress of the iterations. The scalar Report is used to specify this: the condition on Line 15 will be false if Report has the value 0, and true otherwise; on Line 6 it was given the value 1, which is a convenient value to represent true. We can also start this report by printing the initial estimate of the circle's centre. This is done using the statements on Lines 16 to 20.

Two options of PRINT are particularly useful for obtaining a sensible tabular layout to the iteration report. The IPRINT option controls the printing of identifiers; here it has been set to the missing value, to suppress them. The SQUASH option has been used to suppress any blank lines. The third parameter of PRINT, which specifies the number of decimal places to use, has been set to 0 for the iteration counter, and to * for the initial estimate of the centre; Genstat will choose suitable numbers of decimal places, depending on the magnitude of the values. Note that the CHANNEL and IPRINT option settings have carried over from the first PRINT statement to the second, because ampersand (&) has been used to repeat the directive name, and the options.

8.4 The iterative process: fitting the circle

The second part of the program finds the best-fitting circle using a repetitive process. Before showing this part of the program we shall describe how repetitions are specified in Genstat to control iterative processes like the one needed for the circle-fitting program.

The iterations are controlled by the statements shown below.

```
FOR [NTIMES = Lots]
  ...
EXIT Crit < Thresh
  ...
ENDFOR
```

The NTIMES option of the FOR statement is set to the scalar Lots and specifies that the loop is to be executed (at most) that many times, here 100. This puts a maximum on the number of iterations, and means that the Genstat program cannot go into an infinite loop. In fact it is unlikely that all the repetitions will be done, because of the

EXIT statement which is described below. Often it is necessary to have a scalar that holds the iteration number; this could be specified with the statement

```
FOR Iterno = 1...Lots
```

which will automatically increment the iteration counter. However, this will set up a lot of unnamed Genstat scalars (storing the numbers 1, 2, up to Lots individually), and is wasteful of space. It is more efficient to use the NTIMES option in the FOR statement and increment the counter explicitly, using a CALCULATE statement, as we have done here.

A key part of the iteration control is the EXIT statement. This specifies that the FOR-ENDFOR loop is to be jumped out of when the condition Crit<Thresh is true; that is, when the iterative process has converged.

The statements for the complete iterative process are given below. You can see that we have indented the statements within the loop, and also between the IF and ENDIF statements. This is not necessary, but it makes the program easier to read and understand.

```
22    FOR [NTIMES=Lots]
23       CALCULATE Iterno = Iterno + 1
24       & Xadj,Yadj = X,Y - Centre[]
25       & R_hat = MEAN(R = SQRT(Xadj**2 + Yadj**2))
26       & Rss = SUM((R - R_hat) ** 2)
27       & [ZDZ=zero] R = (R > 0) / R
28       & Centre[] = SUM(X,Y * (1 - R)) / SUM(1 - R)
29       & Change = ABS(Oldrss - Rss) / (R_hat ** 2)
30       IF Report
31          PRINT [CHANNEL=2; IPRINT=*; SQUASH=yes] \
32             Iterno,Centre[],R_hat,Rss,Change; \
33             FIELDWIDTH=12; DECIMALS=0,(*)5
34       ENDIF
35       EXIT Change < Thresh
36       CALCULATE Oldrss = Rss
37    ENDFOR
```

The calculation of the new estimate of the radius of the circle is done on Line 25, using the variates Xadj and Yadj formed on Line 24. The residual sum of squares is calculated on Line 26; this needs the distance of each point from the estimate of the centre of the circle, held in the variate R. These values are formed on Line 25, using the expression

```
R = SQRT(Xadj ** 2 + Yadj ** 2)
```

as part of the calculation of R_hat. Because the new estimate of the radius is simply the mean of the values of R, and because they themselves are needed, it is convenient to combine the formation of R and the calculation of R_hat. This will work because in expressions in Genstat the operator = does not just assign a result to a structure; it also leaves the result available for other operators, or a function as in Line 25. However, the = operator is done after all other operators, and in order from left to right if there are several of them. So it is usually necessary to use brackets to force the assignments to take place in the correct order.

To calculate a new estimate of the centre of the circle, in Line 28, we need a variate giving 1/R; this is calculated on Line 27, using a safety device to avoid problems if any of the values of R are zero. The expression (R>0) will give the value 1 for any non-zero values of R; however, for a zero value of R the expression will give 0, so that the calculation will be of 0/0. By setting the ZDZ option of CALCULATE to 'zero' we can ensure that 0/0 will result in 0, which is what is needed by the algebra, rather than a missing value, which is the default result of 0/0.

As explained above, the criterion used to check for convergence is the change in the residual sum of squares. This is calculated on Line 29; however, to make the convergence check independent of the scale of the data values, the change in the residual sum of squares is expressed relative to the squared estimate of the circle's radius.

If the iterative process has not converged then we need to update the scalar holding the residual sum of squares from the last iteration, after the EXIT statement, ready for the next iteration. This is done on Line 36.

The output printed to the secondary channel is shown below. You can see that the number of decimal places for reporting the criterion value has changed as the value has become smaller. This is because the DECIMALS parameter of the PRINT statement for this structure has been set to *, as described above, so that Genstat will use an appropriate number of decimal places.

```
  iteration        A_hat        B_hat        R_hat     Residual    Criterion

          0      0.09640       -2.766
          1      0.09644       -2.715        3.821       0.6883        18.30
          2      0.09622       -2.725        3.823       0.5294      0.01087
          3      0.09630       -2.723        3.823       0.5572     0.001899
          4      0.09628       -2.724        3.823       0.5518    0.0003676
          5      0.09628       -2.724        3.823       0.5528   0.00006980

 Algorithm converged after    5   iterations
```

The EXIT directive is used to jump out of control structures such as FOR-ENDFOR loops and IF-ELSE-ENDIF constructions. It can be used to jump out of any type of control structure, including procedures, but is most generally used to jump out of loops. The second option, CONTROL, of the EXIT directive specifies the type of control structure to exit; by default it is a loop, so it has not been necessary to set the option in this example. It is possible to jump out of several depths of nesting of control structures by setting the first option, NTIMES, of EXIT to specify the depth. The third option, REPEAT, of EXIT is relevant to loops and specifies whether the loop is to be exited completely, or just exited on this pass and re-entered for the next pass. For iterative schemes the former, default, action is appropriate.

8.5 Conclusion: reporting the results

The final part of the program provides a printed report of the iterative process, and optionally draws a graph of the stone ring and fitted circle. The first part of this is shown below.

```
38      FOR
39         IF Change < Thresh
40            PRINT [CHANNEL=2; IPRINT=*] \
41               'Algorithm converged after',Iterno,' iterations'; \
42               FIELDWIDTH=4; DECIMALS=0
43            PRINT [CHANNEL=2] 'Final solution -'
44            & Centre[],R_hat
45            & [IPRINT=*] 'with residual sum of squares',Rss
46         ELSIF Report
47            PRINT [CHANNEL=2] '*** Algorithm failed to converge ***'
48         ELSE
49            PRINT '***  Algorithm failed to converge   ***'
50            & Iterno,Centre[],R_hat,Rss,Change; \
51               FIELDWIDTH=12; DECIMALS=0,(*)5
52         ENDIF
```

When printing the results, the program needs to distinguish between successful completion of the iterative process and any failure to converge. This is tested for on Line 39 and if the program successfully found a solution the PRINT statements on Lines 40 to 45 are executed to report convergence and print the final solution.

The situation for failure-to-converge (Lines 46 to 52) is different. Now it is necessary to draw the user's attention to the failure and it is important to put a message in the appropriate output file. If the user has requested that the progress of the iterative scheme is to be reported to a secondary output file, which is tested for on Line 46, then it will be sufficient to put the failure message after the report of the iterations, as is specified on Line 47. The message includes asterisks to draw attention to it.

The remaining possibility is that the process has not converged and the user has not requested a printed report; in which case the secondary file may be ignored entirely. Now (Lines 49 to 51) it is important that the failure message be put in the primary output file. It is probably also useful for the user to be informed of where the algorithm reached (Lines 50 and 51); this is necessary here only because the information would be at the end of the progress report if one had been requested. The final state of the iterative scheme could be useful because it is quite possible that the estimates obtained are adequate, for example if the convergence threshold had been set too fine.

Not only is it important that the failure message be put in the primary output file, it is also necessary that the printed output from Lines 49 to 51 should not be buried in the middle of a series of Genstat statements, but should appear at the end of the program. Normally Genstat will execute statements as soon as it has read them, so we need to take special action to suppress this. The easiest way is to use a FOR-ENDFOR loop to surround the section of program that we want to be read, but to be executed only when the matching ENDFOR has been reached. The FOR statement is on Line 38, above; the matching ENDFOR (Line 75) does not occur until just before the STOP statement at the end of the program. The FOR statement has no parameters, and no options set, so it will be executed the default number of times, once as required. Also by default, it will compile all the statements before executing any of them, as we need here; however, you could alter this if you wanted by setting the option COMPILE = each.

For the data considered here the iterative scheme converged very rapidly: the report printed to the secondary output file is shown below.

```
Final solution -


       A_hat          B_hat          R_hat
     0.09628        -2.724         3.823



with residual sum of squares        0.5528
```

The last part of the program, to draw a graph, is as follows:

```
53        IF (Change < Thresh) * Graphopt
54           VARIATE [VALUES=0,15...360] Angle
55           CALCULATE Circlex = COS(Angle*6.28318/360)*R_hat+A_hat
56           & Circley = SIN(Angle*6.28318/360)*R_hat+B_hat
57           CASE Graphopt
58              GRAPH [CHANNEL=2; TITLE='Points and fitted circle'; \
59                 JOIN=given; EQUAL=scale; NROWS=37; \
60                 NCOLUMNS=61] B_hat,Circley,Y; A_hat,Circlex,X; \
61                 METHOD=point,curve; SYMBOLS='+',*,'S'
62           OR
63              OPEN 'Circle.grd'; CHANNEL=1; FILETYPE=graphics
64              FRAME 1; YLOWER=0; YUPPER=1; XLOWER=0; XUPPER=1
65              AXES [EQUAL=scale] 1; STYLE=none
66              PEN 1...3; COLOUR=1; METHOD=point,closed; \
67                 LINESTYLE=0,1; SYMBOLS=1,0,4; JOIN=given
68              DGRAPH [TITLE='Points and fitted circle'; WINDOW=1; \
69                 KEYWINDOW=0] B_hat,Circley,Y; A_hat,Circlex,X; \
70                 PEN=1...3
71           ELSE
72              PRINT [IPRINT=*] 'Invalid value of Graphopt',Graphopt
73           ENDCASE
74        ENDIF
75     ENDFOR
76     STOP
```

The intended use of the scalar Graphopt is to specify what type of graph should be drawn: 0 suppresses a graph, 1 specifies a line-printer graph, and 2 specifies a graph drawn using the high-resolution Genstat graphics. We have also decided not to draw a graph unless the iterations converged. Suppression of the graph is tested for on Line 53: the expression within brackets will give the value 1 if the process has converged (as on Line 39); but if Graphopt is zero the complete expression will be 0 and taken as false, so Lines 54 to 73 will be ignored. You can see that we have multiplied the two basic conditions together to check that they are both non-zero:

that is, "true". We could have used the operator .AND. rather than asterisk (*), but the latter saves typing and gives the same result.

The selection of which type of graph to draw is controlled by the CASE, OR, ELSE, and ENDCASE statements (Lines 57, 62, 71, and 73). It might be decided to add different types of graph to those provided, for example a high-resolution graph but with the stones represented by diagrams, rather than single symbols. This would be easy to do by adding an extra OR statement, followed by the statements to draw the new type of graph, between Lines 70 and 71. Now Graphopt would take the value 3 to specify this; however, the PRINT statement after the ELSE statement would still be correct.

There is no difficulty about nesting control structures such as IF-ELSE-ENDIF and FOR-ENDFOR, as shown in the statements that print and graph the results. Because each statement such as IF and FOR must be matched by a corresponding ENDIF or ENDFOR, Genstat can easily work out what the nesting is. However, it is important that you remember these matching statements so that you do not leave one out. If you need to jump out of nested control structures you can use the CONTROL option of the EXIT statement to specify which type of control structure you want to exit.

The fitted circle is drawn using the variates Circlex and Circley giving 25 points around it. For the line-printer graph (Lines 58 to 61) several options need to be set. To ensure that the points for the circle are connected in the correct order the JOIN = given option setting of GRAPH is used. To get a circle drawn, rather than an ellipse, it is necessary to make the physical and mathematical scaling of the x- and y-axes equal. This is as we described in Chapter 5, Section 5.3: a physically square graph is obtained with appropriate settings of the NROWS and NCOLUMNS options of GRAPH, and the mathematical squareness is obtained by the option setting EQUAL = scale. The centre of the circle is marked with the symbol ' + '; default symbols, a mixture of . and ', are used to draw the fitted circle; and the positions of the stones are marked by the letter S.

As mentioned in *Genstat 5: an introduction*, the high-resolution graphical output is controlled in a very different way from the line-printer graphics. First a graphics output file is opened (Line 63). The FRAME statement specifies that Window 1 is to occupy the full graphical area: that is, from 0 to 1 in both x- and y-directions. Thus the output will be square. In the AXES statement it is necessary to set the option EQUAL = scale, analogous to the setting in the GRAPH statement; the STYLE parameter has been set to 'none' to suppress the drawing of any axes, which are irrelevant here. Three pens are used to draw the three parts of the graph: the centre, the fitted circle, and the points for the stones. All three pens have Colour 1 (black); the first and third need to draw points, using symbols coded as 1 and 4, to specify a cross and asterisks, respectively. The method for the second pen needs to be a closed curve, without any symbols, and drawn with Linestyle 1, that is as a full line; the

JOIN parameter, analogous to the JOIN option of the GRAPH statement, specifies that the points are to be joined in the order given. In the DGRAPH statement the option setting KEYWINDOW = 0 is used to suppress the output of a key. Figure 8.2 shows the resulting graph.

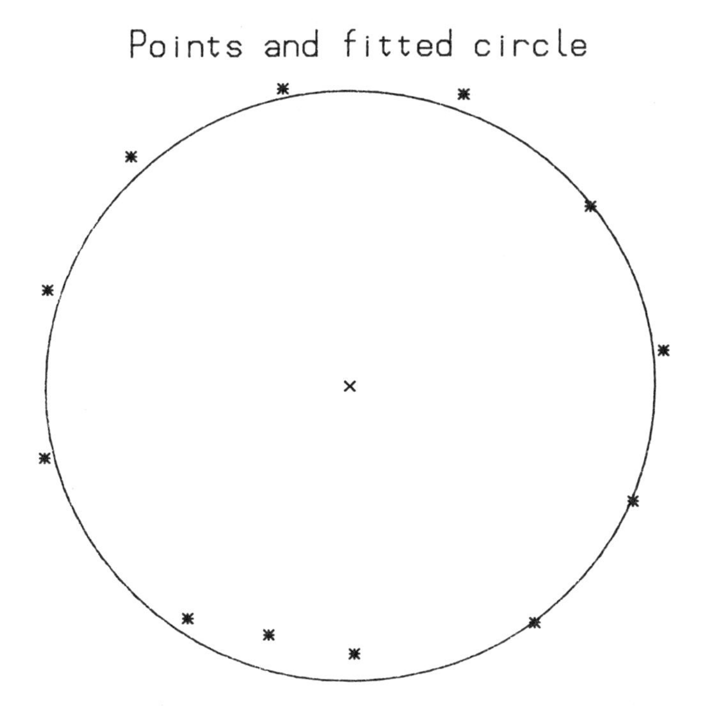

Figure 8.2. High-resolution graph of stone ring and fitted circle.

8.6 Exercises

8(1) Modify the program given in this chapter to allow the user to select between least-squares estimation, as has been used, and least-squares-squared estimation, as suggested by Berman and Culpin (1986). Compare the estimates from the two methods.

8(2) Write a general program to read a set of variates of the same length from a secondary input file. You should assume that the data file contains a scalar giving the number of variates, but that their length is not given. The program should calculate and print several summary statistics for the variates, for example their means and variances, and also produce histograms for each of the variates separately. The production of summary statistics and histograms should be selectable by the user. To test the program you can use many of the data sets used in this book, or a data set of your own.

8(3) Modify the program from Exercise 8(2) to allow the user to specify that graphs should also be drawn of all the pairs of the variates; for example, for three variates, V1, V2, and V3, plot V2 against V1, V3 against V1, and V3 against V2. The program should be able to do this for any number of variates. This is quite difficult; however, it is a useful facility to know how to program.

9 Writing procedures in the Genstat language

9.1 The need for new Genstat commands

Most regular users of Genstat eventually feel a need for some commands that the language does not provide, though different users have different commands on their "shopping list". Each of their requirements can usually be met by a sequence of Genstat statements, just as it is possible to perform an analysis of variance using FOR loops containing RESTRICT and CALCULATE statements, as if there were no BLOCKS, TREATMENTS, and ANOVA directives. Such a sequence can be placed in each program wherever it is required, with *ad hoc* modifications; but such modifications require a detailed knowledge of the working of the sequence of statements, and carry a risk of error or omission. A more economical and elegant solution is to place the sequence of statements in a named *procedure* which can then be used like other Genstat directives, complete with options and parameters to suit varying circumstances.

In this chapter we shall describe how you can design and implement your own procedures, and also how to gain access to those written by others. In order to write such generalized programs, it will be necessary to use Genstat more as a programming language and less as a statistical toolkit than in earlier examples.

9.2 The Kruskal–Wallis analysis of variance of ranks

A commonly used statistical technique that is not available through the standard Genstat directives is the Kruskal–Wallis one-way analysis of variance of ranks. This differs from the usual one-way analysis of variance in that it is based on the ranks of the data values (highest, second highest, third highest, and so on) rather than on the values themselves. Consequently it assesses differences between medians rather than means, and its validity is not dependent upon the residual variation being Normally distributed. Because of this robustness it is often recommended to investigators with little knowledge of statistics. Such investigators may also find it more natural to arrange their data with the treatments in different rows, rather than with two columns containing values of a factor and a variate. This way of arranging the data can be illustrated using an example given by Boer and Jansen (1942). Rats were fed with four subdiets of a diet A, designated A_1, A_2, A_3, and A_4, in order to determine whether the subdiets could be regarded as equivalent. The growth values obtained, in standard units, are arranged below for reading by Genstat, with the rows corresponding to the subdiets. The initial 4 indicates the number of treatments.

```
4:
257   205   206   164   190   214   228   203:
201   231   197   185:
248   265   187   220   212   215   281:
202   276   207   206   230   227:
```

We shall first produce a Genstat program that will read the data in this form, concatenate the rows into a variate and produce an accompanying factor, and perform a Kruskal–Wallis analysis on these. We shall then consider how to make this program into a procedure.

The following statements will read, concatenate, and print the data.

```
SCALAR Ntreat
OPEN 'Ratser.dat'; CHANNEL = 2
READ [CHANNEL = 2] Ntreat
& [SERIAL = yes; SETNVALUES = yes] A[1...Ntreat]
CALCULATE N[1...Ntreat] = NVALUES(A[])
FOR I = 1...Ntreat
    VARIATE [VALUES = (I)#N[I]] V[I]
ENDFOR
FACTOR [LEVELS = Ntreat; VALUES = #V[]] Subdiet
VARIATE [VALUES = #A[]] Growth
PRINT Subdiet,Growth
```

The scalar Ntreat is used to hold the number of variates to be concatenated, and the scalars N[1...Ntreat] are used to hold the number of values of each of the variates A[1...Ntreat]. Since A is automatically declared as a pointer whose values are A[1...Ntreat] this list of identifiers can be succinctly represented by A[]. Thus the CALCULATE statement places the number of values in A[1] into N[1], the number of values in A[2] into N[2], and so on. Note the difference between NVALUES(A[]) and NVALUES(A), which would give simply the number of values of the pointer A, in this case 4. Note also that when the scalars N[1...Ntreat] are first referred to, in the CALCULATE statement, they cannot be represented by N[].

The accompanying factor is created by first creating separate variates V[1...Ntreat] for each treatment. The hash symbol (#) has the effect of "unpacking" a structure, as if its values, rather than its identifier, had been given. (It is the inverse of the exclamation mark (!) which "bundles" a list of values into an unnamed structure.) Thus when I stands for 1, #N[I] stands for the value of N[1], which is 8, so that (I)#N[I] stands for (1)8. Thus V[1] is of length 8 and all its values are 1, and so on. A factor Subdiet is declared, and the values of V[1...Ntreat], which can be represented by #V[1...Ntreat] and hence by #V[], are placed in it. Similarly, the values of A[1...Ntreat] are placed in the variate Growth. Subdiet and Growth are then printed.

The following statements replace the values of Growth by their ranks:

```
SORT [INDEX = Growth; GROUPS = Frank]
CALCULATE Growth = Frank
```

The SORT statement obtains the ranks of the index variate Growth, and places them in the factor Frank, each rank being a different level of this factor. The CALCULATE statement puts the ranks back into the variate Growth.

The test statistic for the significance of variation among the treatment medians is given by

$$K = \left(\frac{12}{N(N+1)} \sum_{i=1}^{m} \frac{\left(\sum_{j=1}^{n_i} R_{ij} \right)^2}{N_i} \right) - 3(N+1)$$

where:

R_{ij} = rank of the jth value for the ith treatment,
N_i = number of values for the ith treatment, and
m = number of treatments,

$$N = \sum_{i=1}^{m} N_i = \text{total number of values.}$$

This statistic is calculated by the following statements, which also create tables to retain the number of values and the mean rank in each treatment.

```
TABULATE [CLASSIFICATION = Subdiet; COUNT = Counttab] \
    Growth; TOTALS = Totrank; MEANS = Meantab
CALCULATE Ntot = NVALUES(Growth)
& K = 12 / (Ntot * (Ntot + 1)) * \
    SUM(Totrank ** 2 / Counttab) - 3 *(Ntot + 1)
```

The following statements print the results of the analysis, with some appropriate text.

```
PRINT 'Kruskal–Wallis One-way Anova'
& [ORIENTATION = across; IPRINT = *] 'Value of K = ',K
& 'Sample size:',Counttab; DECIMALS = 0
& 'Mean ranks:',Meantab
```

```
FOR I = 1...Ntreat
    RESTRICT Growth; Subdiet .EQ. I
    PRINT [ORIENTATION = across; IPRINT = *] 'Ranks of treatment',I; \
        DECIMALS = 0
    & [ORIENTATION = down] Growth; DECIMALS = 0
ENDFOR
STOP
```

The ranks for each treatment need to be preceded by a heading giving the treatment number. An appropriate text and the number are printed across the page, and the ranks are printed below. Printing of the identifiers of these structures is suppressed by using an asterisk as the setting of the option IPRINT. The asterisk is used as a setting of many other options in this way, by analogy with its use to represent a missing value in data.

9.3 Creating and using a procedure

In deciding how to make these statements into a procedure, it is necessary to consider how the procedure will be used in other Genstat programs. It should be obtainable by means of statements such as:

```
SCALAR Ntreat
OPEN 'Ratser.dat'; CHANNEL = 2
READ [CHANNEL = 2] Ntreat
& [SERIAL = yes; SETNVALUES = yes] A[1...Ntreat]
KRUSKAL VARIATES = A
```

The Genstat program within the procedure KRUSKAL must begin with a series of statements that define the syntax of the statement that calls it. The first of these is simply

```
PROCEDURE 'KRUSKAL'
```

which allows you to use KRUSKAL in the rest of the program as if it were a Genstat directive name. This is followed by

```
PARAMETER NAME = 'VARIATES'; MODE = p
```

This indicates that future KRUSKAL statements can have a parameter named VARIATES. This name will also be used as an identifier within the procedure: the name must be in capitals within the procedure, though when used as a parameter to invoke the procedure it may be in upper or lower case. The parameter is of mode 'p' because its setting is a list of identifiers, as are the values of a pointer. Note that the PARAMETER directive itself has two parameters, NAME and MODE. In any computer language in which it is possible, the use of commands that define other commands requires clear thinking!

Within the procedure, each occurrence of the identifier A must be replaced by VARIATES. By extension, #A must be replaced by #VARIATES, and #A[] by #VARIATES[]. The variates specified by the VARIATES parameter will no longer necessarily represent growth, and Growth is therefore no longer an appropriate name for the variate obtained by concatenating them. Each occurrence of Growth will therefore be replaced by Response. Similarly the factor that distinguishes the groups of values will no longer necessarily be Subdiet, and it will simply be called FACTOR. (The reason for choosing this name and writing it in capitals will become apparent in Section 9.4.)

Although the number of treatments was read from the data, this value is not automatically available within the procedure because it was not used as a parameter value when the procedure was called. It must therefore be calculated by the statement:

```
CALCULATE Ntreat = NVALUES(VARIATES)
```

Note that there are now two distinct structures called Ntreat. This does not cause ambiguity since one is available only in the main program and the other only in the procedure.

The procedure is ended by the statement:

```
ENDPROCEDURE
```

The complete program is shown below. Note that the procedure comes first, followed by the statements that call it. Note also that an explanatory comment has been included at the beginning of the procedure.

```
PROCEDURE 'KRUSKAL'
PARAMETER NAME = 'VARIATES'; MODE = p
"This procedure conducts the Kruskal–Wallis analysis of variance
of ranks. The VARIATES parameter supplies the observations,
each treatment in a separate variate."
CALCULATE Ntreat = NVALUES(VARIATES)
CALCULATE N[1...Ntreat] = NVALUES(VARIATES[])
FOR I = 1...Ntreat
    VARIATE [VALUES = (I)#N[I]] V[I]
ENDFOR
FACTOR [LEVELS = Ntreat; VALUES = #V[]] FACTOR
VARIATE [VALUES = #VARIATES[]] Response
PRINT FACTOR,Response
SORT [INDEX = Response; GROUPS = Frank]
CALCULATE Response = Frank
```

```
TABULATE [CLASSIFICATION = FACTOR; COUNT = Counttab] \
    Response; TOTALS = Totrank; MEANS = Meantab
CALCULATE Ntot = NVALUES(Response)
& K = 12 / (Ntot * (Ntot + 1)) * \
    SUM(Totrank ** 2 / Counttab) - 3 * (Ntot + 1)
PRINT 'Kruskal-Wallis One-way Anova'
& [ORIENTATION = across; IPRINT = *] 'Value of K = ',K
& 'Sample size:',Counttab; DECIMALS = 0
& 'Mean ranks:',Meantab
FOR I = 1...Ntreat
    RESTRICT Response; FACTOR .EQ. I
    PRINT [ORIENTATION = across; IPRINT = *] \
        'Ranks of treatment',I; DECIMALS = 0
    & [ORIENTATION = down] Response; DECIMALS = 0
ENDFOR
ENDPROCEDURE
SCALAR Ntreat
OPEN 'Ratser.dat'; CHANNEL = 2
READ [CHANNEL = 2] Ntreat
& [SERIAL = yes; SETNVALUES = yes] A[1...Ntreat]
KRUSKAL VARIATES = A
STOP
```

9.4 Making a procedure more flexible

As it stands, the procedure KRUSKAL is less flexible than most Genstat directives, because it offers the user no control over its output. It would also be useful to be able to choose whether to use a separate variate for each treatment, or to use a single variate and specify the treatments in a factor. Some additions to the syntax of the KRUSKAL procedure, and changes within the procedure, will make these choices available.

The output will be controlled by a PRINT option with settings 'stats' and 'data'. The default setting will be 'stats' alone, and all printing will be suppressed by a third setting, 'nothing'. It is possible to arrange that the setting PRINT = * also suppresses output, as with all standard Genstat directives that have a PRINT option. However, this would need some extra statements in the procedure to avoid problems caused by the dummy called PRINT pointing to a structure with no values.

Another option, TRTSPEC, will indicate whether the treatments are specified by separate variates or by a factor. Its settings will be 'rows' and 'factor' respectively, with 'rows' as the default setting. When TRTSPEC is set to 'factor', a second parameter, FACTOR, will be needed to indicate where the treatments are specified.

Thus more elaborate KRUSKAL statements will be possible, as in the following example:

```
UNITS [NVALUES = 25]
VARIATE Growth
FACTOR [LABELS = !T(A1,A2,A3,A4)] Subdiet
OPEN 'Ratpar.dat'; CHANNEL = 2
READ [CHANNEL = 2] Subdiet,Growth
KRUSKAL [TRTSPEC = factor; PRINT = stats,data] VARIATES = Growth; \
    FACTOR = Subdiet
```

In order to define the options, the following statement is added after the PROCEDURE and PARAMETER statements:

```
OPTION NAME = 'TRTSPEC','PRINT'; MODE = t; DEFAULT = 'rows','stats'
```

The option names TRTSPEC and PRINT will be used as identifiers within the procedure, and the parameter MODE indicates the type of structure they identify. In this case they are text structures (mode 't'), since their values—the option settings—are strings of alphanumeric characters. Since both options have the same mode, this is given once, and the single-item list is recycled. Option settings can also be numerical values, in which case the option name is treated as a variate (mode 'v'), or identifiers, in which case the option name is treated as a pointer (mode 'p'). In mode 't', the difference between lower and upper case is significant, and hence the settings 'rows' and 'Rows' are not equivalent.

The new parameter FACTOR is added to the PARAMETER statement:

```
PARAMETER NAME = 'VARIATES','FACTOR'; MODE = p
```

The action of the statements within the procedure must now be controlled by IF statements that check the settings of the parameter FACTOR and of the options. Within KRUSKAL, the parameter name FACTOR is treated as the identifier of a pointer. If it has not been set when KRUSKAL is called, then some other identifier must be assigned to it so that subsequent references to FACTOR point to a definite structure. This is achieved by the following statements:

```
IF UNSET(FACTOR)
    ASSIGN Dfactor; POINTER = FACTOR
ENDIF
```

The parameter VARIATES should never be unset when the procedure is called, and we shall not attempt to make our procedure "user friendly" by providing for this eventuality.

The concatenation of variates and the creation of the accompanying factor need take place only if 'factor' is not in the setting of the option TRTSPEC when the procedure KRUSKAL is called. Otherwise it is necessary only to copy the values from the single variate specified by the VARIATES parameter into Response. Thus the concatenation statements are preceded by

```
IF 'factor' .NI. TRTSPEC
```

and followed by

```
ELSE
    VARIATE [VALUES = #VARIATES] Response
ENDIF
```

The operator .NI. stands for "Not In".

The statistics should be printed only if 'stats' is included in the setting of the PRINT option, and the statements

```
PRINT 'Kruskal–Wallis One-way Anova'
& [ORIENTATION = across; IPRINT = *] 'Value of K = ',K
& 'Sample size:',Counttab; DECIMALS = 0
& 'Mean ranks:',Meantab
```

are therefore preceded by

```
IF 'stats' .IN. PRINT
```

and followed by

```
ENDIF
```

Similarly, the statements that print the ranks will be preceded by

```
IF 'data' .IN. PRINT
```

and followed by

```
ENDIF
```

In order to make the form in which the ranks are printed appropriate to the setting of TRTSPEC, the statements within this IF-ENDIF block will be preceded by

```
IF 'factor' .NI. TRTSPEC
```

and followed by

```
ELSE
    PRINT FACTOR,Response
ENDIF
```

The complete procedure is now as follows. Note that the comment has been suitably modified.

```
PROCEDURE 'KRUSKAL'
OPTION NAME = 'TRTSPEC','PRINT'; MODE = t; DEFAULT = 'rows','stats'
PARAMETER NAME = 'VARIATES','FACTOR'; MODE = p
"This procedure conducts the Kruskal–Wallis analysis of variance
of ranks. The VARIATES parameter supplies the observations,
either with each treatment in a different variate or with all
in one variate. In the latter case they are classified by a
factor, supplied by the FACTOR parameter. The TRTSPEC option
indicates the organization of the data and is set to 'rows'
(for several variates) or 'factor' (for one variate and a factor).
The PRINT option controls printing and is set to some
combination of 'stats' and 'data'."
IF UNSET(FACTOR)
    ASSIGN Factor; POINTER = FACTOR
ENDIF
IF 'factor' .NI. TRTSPEC
    CALCULATE Ntreat = NVALUES(VARIATES)
    CALCULATE N[1...Ntreat] = NVALUES(VARIATES[])
    FOR I = 1...Ntreat
        VARIATE [VALUES = (I)#N[I]] V[I]
    ENDFOR
    FACTOR [LEVELS = Ntreat; VALUES = #V[]] FACTOR
    VARIATE [VALUES = #VARIATES[]] Response
    IF 'data' .IN. PRINT
        PRINT FACTOR,Response
    ENDIF
ELSE
    VARIATE [VALUES = #VARIATES] Response
ENDIF
SORT [INDEX = Response; GROUPS = Frank]
CALCULATE Response = Frank
TABULATE [CLASSIFICATION = FACTOR; COUNT = Counttab] \
    Response; TOTALS = Totrank; MEANS = Meantab
CALCULATE Ntot = NVALUES(Response)
& K = 12 / (Ntot * (Ntot + 1)) * \
    SUM(Totrank ** 2 / Counttab) – 3 * (Ntot + 1)
IF 'stats' .IN. PRINT
    PRINT 'Kruskal–Wallis One-way Anova'
    & [ORIENTATION = across; IPRINT = *] 'Value of K = ',K
    & 'Sample size:',Counttab; DECIMALS = 0
    & 'Mean ranks:',Meantab
```

```
ENDIF
IF 'data' .IN. PRINT
    IF 'factor' .NI. TRTSPEC
        FOR I = 1...Ntreat
            RESTRICT Response; FACTOR .EQ. I
            PRINT [ORIENTATION = across; IPRINT = *] \
                'Ranks of treatment',I; DECIMALS = 0
            & [ORIENTATION = down] Response; DECIMALS = 0
        ENDFOR
    ELSE
        PRINT FACTOR,Response
    ENDIF
ENDIF
ENDPROCEDURE
```

9.5 Using a procedure within a procedure

The statements that perform the concatenation process are likely to be useful in other contexts. They can therefore be placed in a procedure to be used within the procedure KRUSKAL. There is a Genstat directive called CONCATENATE, which means that we are prohibited from giving the new procedure this name: instead we shall call it CHAIN. This procedure will have a parameter INVARS, to specify the variates to be concatenated, and parameters OUTVAR and OUTFACT, to specify the output variate and factor respectively. It can also have a PRINT option, so that the concatenated variate and accompanying factor can be printed if required. The PRINT option of CHAIN should clearly have the settings 'yes' and 'no'. However, which of these is chosen by the procedure KRUSKAL should depend on whether 'data' has been included in the setting of the PRINT option of KRUSKAL. This is achieved by the following statements:

```
IF 'factor' .NI. TRTSPEC
    CALCULATE Ntreat = NVALUES(VARIATES)
    IF 'data' .IN. PRINT
        TEXT [VALUES = 'yes'] Printset
    ELSE
        TEXT [VALUES = 'no'] Printset
    ENDIF
    CHAIN [PRINT = #Printset] INVARS = VARIATES; \
        OUTVAR = Response; OUTFACT = FACTOR
ELSE
    VARIATE [VALUES = #VARIATES] Response
ENDIF
```

These replace the concatenation statements within KRUSKAL.

When using CHAIN, it is essential that the variates to be concatenated should be represented by a single pointer: a list of variates would imply that the procedure was to be used several times. Thus:

```
VARIATES X,Y,Z
CHAIN INVARS = X,Y,Z; OUTVAR = Vall; OUTFACT = F
```

is not acceptable. The desired effect can be obtained either by

```
VARIATES X,Y,Z
POINTER [VALUES = X,Y,Z] V
CHAIN INVARS = V; OUTVAR = Vall; OUTFACT = F
```

or more economically by

```
VARIATES X,Y,Z
CHAIN INVARS = !P(X,Y,Z); OUTVAR = Vall; OUTFACT = F
```

Note that the procedure can indeed be used more than once by a single statement, to concatenate several sets of variables, as follows:

```
VARIATES W[1...5],X,Y,Z
CHAIN INVARS = W,!P(X,Y,Z); OUTVAR = Wall,Vall; \
    OUTFACT = F[1,2]
```

The statements defining the procedure CHAIN are as follows:

```
PROCEDURE 'CHAIN'
OPTION NAME = 'PRINT'; MODE = t; DEFAULT = 'no'
PARAMETER NAME = 'INVARS','OUTVAR','OUTFACT'; MODE = p
"This procedure concatenates several variates, supplied by the
INVARS parameter, into one, supplied by the OUTVAR parameter.
The OUTFACT parameter supplies a factor to indicate to which of
the input variates each value in the output variate belonged.
The PRINT option controls printing and is set to 'no' or 'yes'."
CALCULATE Nvar = NVALUES(INVARS)
CALCULATE N[1...Nvar] = NVALUES(INVARS[])
FOR I = 1...Nvar
    VARIATE [VALUES = (I)#N[I]] V[I]
ENDFOR
FACTOR [LEVELS = Nvar; VALUES = #V[]] OUTFACT
VARIATE [VALUES = #INVARS[]] OUTVAR
IF 'yes' .IN. PRINT
    PRINT OUTFACT,OUTVAR
ENDIF
ENDPROCEDURE
```

The default setting of the PRINT option is 'no', as is the case with all options of directives that offer a 'yes' or 'no' choice. The separate variates to be concatenated will not now necessarily represent treatments, so the scalar Ntreat is changed to Nvar. Since Nvar is not used as an option or parameter of CHAIN, its value must be calculated within the procedure. The identifiers VARIATES, Response, and FACTOR are replaced by INVARS, OUTVAR, and OUTFACT respectively.

9.6 Using a Library procedure

So far, no provision has been made for the possibility that the data contain missing values. We shall not attempt to ensure that the procedure KRUSKAL performs a correct analysis when missing values are present, but it should at least check whether there are any and, if so, print a warning. This can be done by a procedure called DESCRIBE, which provides summary statistics about variates and which is available in the Genstat Procedure Library. It can be used by inserting the following statements before the SORT statement in the procedure KRUSKAL:

```
DESCRIBE [SELECT = nmv] Response; SUMMARIES = Nummiss
EXIT [CONTROL = procedure] Nummiss .GT. 0
```

The option setting SELECT = nmv indicates that the number of missing values is the summary statistic to be selected, and the SUMMARIES parameter indicates that its value is to be stored in the variate Nummiss. The EXIT statement ensures that no subsequent statements in the procedure will be executed if there are any missing values. The DESCRIBE procedure will print a message about the number of missing values, so that if any are found, a message such as that following will be printed before exiting from the KRUSKAL procedure.

```
Summary statistics for Response

   Number of missing values = 2
```

DESCRIBE is in the Genstat Procedure Library, which should be available wherever Genstat 5 is installed. Each Genstat site also has its own Site Procedure Library which is available whenever Genstat is used at that site, and you can create your own procedure libraries and arrange for them to be available whenever you use Genstat: we describe how to do this in Section 9.8. If Genstat encounters a statement that does not begin with one of the standard directive names, it looks first for a procedure with that name, defined in the current job. If it cannot find one, it scans any procedure libraries that you have attached using OPEN (see Section 9.8), then the Site Procedure Library, then the Genstat Procedure Library for a procedure with that name. Thus in naming your procedures you do not need to worry that there may be a standard procedure of the same name, though the existence of your procedure will make such a standard procedure unavailable.

9.7 Modular programming

In outline, the complete program now has the modular structure shown in Figure 9.1.

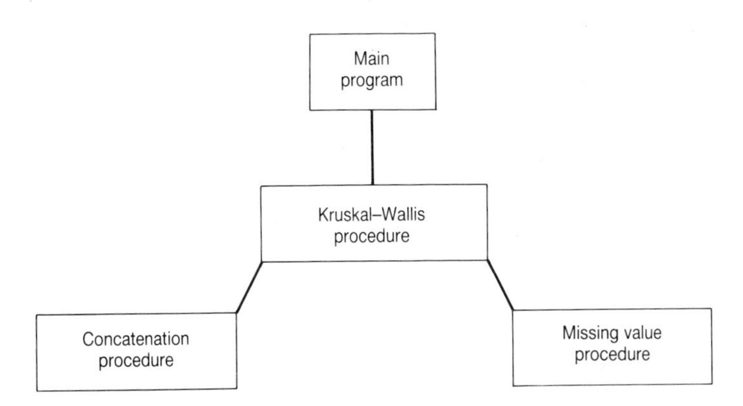

Figure 9.1. Modular structure of complete program.

Each box is called by the box above, and returns control to the box above when it has finished working. In the present example we have started with a single program and progressively broken it into these modules, but it is sometimes easier to start with a diagram like this, and to think about the details from the top down, initially considering only the input required by a box when it is called, and the output that it returns. If it is known beforehand that a procedure is needed then a procedure should be designed from the outset because, as you have seen, the changes needed to turn a program into a procedure are detailed and rather unpredictable. They can, however, be classified as follows:

(1) replacing identifiers by option and parameter names;
(2) changing specific identifiers that are inappropriate within the more general procedure;
(3) changing IF and ELSIF statements to operate on option settings instead of other structure values;
(4) obtaining values that are not brought in from the main program as option or parameter settings;
(5) in the main program, obtaining values that are not made available from the procedure through option or parameter settings.

The output of the completed program, including the listing, is as follows. In the present case, the procedures are not stored in libraries but are in the same file as the main program. Note that CHAIN is and must be listed before KRUSKAL, by which it is called, and likewise KRUSKAL must be listed before the main program.

```
  1  PROCEDURE 'CHAIN'
  2  OPTION NAME='PRINT'; MODE=t; DEFAULT='no'
  3  PARAMETER NAME='INVARS','OUTVAR','OUTFACT'; MODE=p
  4  "This procedure concatenates several variates, supplied by the
 -5   INVARS parameter, into one, supplied by the OUTVAR parameter.
 -6   The OUTFACT parameter supplies a factor to indicate to which of
 -7   the input variates each value in the output variate belonged.
 -8   The PRINT option controls printing and is set to 'no' or 'yes'."
  9  CALCULATE Nvar = NVALUES(INVARS)
 10  CALCULATE N[1...Nvar] = NVALUES(INVARS[])
 11  FOR I=1...Nvar
 12     VARIATE [VALUES=(I)#N[I]] V[I]
 13  ENDFOR
 14  FACTOR [LEVELS=Nvar; VALUES=#V[]] OUTFACT
 15  VARIATE [VALUES=#INVARS[]] OUTVAR
 16  IF 'yes' .IN. PRINT
 17     PRINT OUTFACT,OUTVAR
 18  ENDIF
 19  ENDPROCEDURE
 20
 21  PROCEDURE 'KRUSKAL'
 22  OPTION NAME='TRTSPEC','PRINT'; MODE=t; DEFAULT='rows','stats'
 23  PARAMETER NAME='VARIATES','FACTOR'; MODE=p
 24  "This procedure conducts the Kruskal-Wallis analysis of variance
-25   of ranks. The VARIATES parameter supplies the observations,
-26   either with each treatment in a different variate or with all
-27   in one variate. In the latter case they are classified by a
-28   factor, supplied by the FACTOR parameter. The TRTSPEC option
-29   indicates the organization of the data and is set to rows
-30   (for several variates) or factor (for one variate and a factor).
-31   The PRINT option controls printing and is set to some
-32   combination of 'stats' and 'data'."
 33  IF UNSET(FACTOR)
 34     ASSIGN Factor; POINTER=FACTOR
 35  ENDIF
 36  IF 'factor' .NI. TRTSPEC
 37     CALCULATE Ntreat = NVALUES(VARIATES)
 38     IF 'data' .IN. PRINT
 39        TEXT [VALUES='yes'] Printset
 40     ELSE
 41        TEXT [VALUES='no'] Printset
 42     ENDIF
 43     CHAIN [PRINT=#Printset] INVARS=VARIATES; \
 44        OUTVAR=Response; OUTFACT=FACTOR
 45  ELSE
 46     VARIATE [VALUES=#VARIATES] Response
 47  ENDIF
 48  DESCRIBE [SELECT=nmv] Response; SUMMARIES=Nummiss
 49  EXIT [CONTROL=procedure] Nummiss .GT. 0
 50  SORT [INDEX=Response; GROUPS=Frank]
 51  CALCULATE Response = Frank
 52  TABULATE [CLASSIFICATION=FACTOR; COUNT=Counttab] \
 53     Response; TOTALS=Totrank; MEANS=Meantab
 54  CALCULATE Ntot = NVALUES(Response)
 55  & K = 12 / (Ntot * (Ntot + 1)) * \
 56     SUM(Totrank ** 2 / Counttab) - 3 * (Ntot + 1)
 57  IF 'stats' .IN. PRINT
 58     PRINT 'Kruskal-Wallis One-way Anova'
 59     & [ORIENTATION=across; IPRINT=*] 'Value of K =',K
 60     & 'Sample size:',Counttab; DECIMALS=0
 61     & 'Mean ranks:',Meantab
 62  ENDIF
 63  IF 'data' .IN. PRINT
 64     IF 'factor' .NI. TRTSPEC
 65        FOR I=1...Ntreat
 66           RESTRICT Response; FACTOR .EQ. I
 67           PRINT [ORIENTATION=across; IPRINT=*] \
 68              'Ranks of treatment',I; DECIMALS=0
 69           & [ORIENTATION=down] Response; DECIMALS=0
```

```
70          ENDFOR
71       ELSE
72          PRINT FACTOR,Response
73       ENDIF
74    ENDIF
75    ENDPROCEDURE
76
77    SCALAR Ntreat
78    OPEN 'Ratser.dat'; CHANNEL=2
79    READ [CHANNEL=2] Ntreat
80    READ [CHANNEL=2; SERIAL=yes; SETNVALUES=yes] A[1...Ntreat]

   Identifier    Minimum       Mean    Maximum     Values    Missing
         A[1]      164.0      208.4      257.0          8          0
         A[2]      185.0      203.5      231.0          4          0
         A[3]      187.0      232.6      281.0          7          0
         A[4]      202.0      224.3      276.0          6          0
81    FACTOR F
82    KRUSKAL VARIATES=A; FACTOR=F


Summary statistics for Response

   Number of missing values = 0


Kruskal-Wallis One-way Anova

Value of K =
       4.213

Sample size:

           F
           1           8
           2           4
           3           7
           4           6

Mean ranks:

           F
           1       11.00
           2        8.25
           3       16.57
           4       14.67

83
84    STOP
```

The test statistic K is distributed approximately as χ^2 with $m - 1$ degrees of freedom where m is the number of treatments. From Table 8 of *New Cambridge elementary statistical tables* (Lindley and Scott 1984) it is found that the observed value, K = 4.2, falls far short of the 5% critical value with three degrees of freedom, which is 7.815. In fact it is below the 20% critical value, 4.642. It is therefore probably safe to pool the subdiets A_1 to A_4.

This complete program has worked satisfactorily, but in order to test the procedures KRUSKAL and CHAIN fully it would be necessary to use them with each possible combination of option and parameter settings. When a procedure has several options each with several settings, such an exhaustive test may be impractical.

Another limitation of the procedures as they stand is that there is no checking within the procedures for errors or ambiguities when they are called. For example, if the KRUSKAL procedure is called with the option setting PRINT = stars,data, no statistics would be printed: there is no check within KRUSKAL for invalid settings of the PRINT option. Such checks can be incorporated using suitable IF, ELSE, and ENDIF statements, but it is up to you to decide how rigorous it is worth making such checks.

9.8 Storing procedures for future use

Procedures may occasionally be used simply to give a modular structure to a particular program, but more often they are required to be stored for use by a variety of programs. Like variates, factors, and so on, procedures are Genstat structures and their names can be treated as identifiers. As an alternative to reading the values of a Genstat structure from an ordinary file on your computer system, it is also possible to hold a structure, complete with its identifier, attributes, and values, in a special backing-store file whose internal layout is arranged by the computer. A library of procedures can thus be built and edited in such a file using Genstat statements.

Such backing-store files can be used to make procedures available to many programs. For example, after the procedures KRUSKAL and CHAIN have been created using the directives PROCEDURE, ENDPROCEDURE, and so on, they can both be stored in the file Myprocs.dat by the statements:

```
OPEN 'Myprocs.dat'; CHANNEL = 2; FILETYPE = backingstore
STORE [CHANNEL = 2; PROCEDURE = yes] KRUSKAL,CHAIN
```

The OPEN statement opens the file Myprocs.dat and connects it to backing-store Channel 2. The setting of the FILETYPE parameter identifies which type of file is being connected: the default, as used often in previous chapters, is 'input' as required for files to be accessed by READ statements. Backing-store files contain whole Genstat structures, including procedures, recording information about structures as well as their values. The channel numbers of different file types are independent: thus there might also be a file containing data, connected to input Channel 2. The STORE statement causes the procedure structures KRUSKAL and CHAIN to be stored in Myprocs.dat. The PROCEDURE option specifies that the identifiers in the STORE statement are names of procedures rather than of other data structures: in fact, you cannot mix procedures with other structures in a STORE statement.

The backing-store file must be given a name that is valid on the computer system being used, and thus the OPEN directive is one of the few in which the syntactic validity of a Genstat statement depends on the system outside. For example, the statement

```
OPEN NAME = '.Myprocs:dat'; CHANNEL = 1; FILETYPE = backingstore
```

would usually be valid on an IBM 3081 computer system but on a VAX 11/750 with the operating system VMS 4 it would produce the following error:

```
******** Fault (Code IO 22). Statement 1 on Line 78
Command: OPEN NAME='.Myprocs:dat'; CHANNEL=2; FILETYPE=backingstore

Fortran error when opening or closing a file
Backingstore File on Channel 2  IOSTAT = 43

A fatal fault has occurred - the rest of this job will be ignored
```

The proccdures can be used in a subsequent program by the statements:

```
OPEN NAME = 'Myprocs.dat'; CHANNEL = 3; FILETYPE = procedure
SCALAR Ntreat
OPEN 'Ratser.dat'; CHANNEL = 2; FILETYPE = input
READ [CHANNEL = 2] Ntreat
& [SERIAL = yes; SETNVALUES = yes] A[1...Ntreat]
KRUSKAL VARIATES = A
```

The OPEN statement connects the file Myprocs.dat through Channel 3 (Channel 2 could have been used again). Note that the FILETYPE parameter is now set to 'procedure' rather than to 'backingstore': this causes Genstat to search this file automatically whenever an unknown directive name is found. The procedures can then be used.

If you define a particular file to be the Site Procedure Library, you can use the procedures that it contains without opening it. However, the method of defining this Library varies from site to site and you will have to consult your local advisers about this.

A procedure library file can be divided into subfiles and you can add new procedures to your library by placing them in new subfiles. Thus the statement

```
STORE [CHANNEL = 1; SUBFILE = Nonpars; METHOD = overwrite; \
    PROCEDURE = yes] KRUSKAL,CHAIN
```

will cause the structures KRUSKAL and CHAIN to be placed in the subfile Nonpars of the file connected to Channel 1. The option METHOD = overwrite specifies that this subfile is to overwrite any existing subfile named Nonpars. The default setting is 'add', which would give an error message if a subfile named Nonpars already existed.

9.9 Exercises

9(1) Convert the series of statements that performs Bartlett's test for homogeneity of variance, given in Chapter 1, Section 1.3, into a procedure. The procedure should require a series of sums of squares and a series of corresponding degrees of freedom, each series stored in a variate. An option should specify whether the test statistic and its degrees of freedom are to be printed. They should be printed by default, and it should also be possible to return them to the main program as scalars.

Write statements to test the procedure:

(a) with the printing of values specified explicitly,
(b) with the printing of values obtained by default, and
(c) with no printing done by the procedure, but with the values returned to the main program and printed.

9(2) Write a procedure to convert times of day into minutes or seconds after midnight and return the converted values to the main program. An option should specify whether the input is given in hours and minutes or in hours, minutes, and seconds, and this should determine whether the output is given in minutes or seconds. Another option should specify whether the input is given in 12-hour or 24-hour clock notation. If the 12-hour notation is used, the input will include a factor with levels labelled AM and PM.

Write statements to test the procedure:

(a) with times not including seconds specified in the 12-hour clock notation, and
(b) with times including seconds specified in the 24-hour clock notation.

9(3) The statements below are part of a procedure for determining sowing dates of crops. The mean temperature on a series of dates starting from the first of January is provided and the statements calculate, at each date, the cumulative mean difference between the temperature and a critical value T. For temperatures below T, their difference from T either may have a negative value or may be treated as zero. The number of periods during which the temperature exceeded T and their total duration may also be calculated. The differences and number and duration of periods may be printed, and additionally may be returned to the main program as a variate and two scalars.

```
UNITS TEMP
CALCULATE Tempdiff = TEMP - CRITT
IF 'yes' .IN. NEGSZERO
    CALCULATE Tempdiff = Tempdiff * (Tempdiff .GT. 0)
ENDIF
CALCULATE Lastday = NVALUES(TEMP)
VARIATE [VALUES = 1...Lastday] Day
CALCULATE MEANDIFF = CUMULATE(Tempdiff) / Day
CALCULATE Exceed = Tempdiff .GT. 0
IF 'yes' .IN. NUMBER
```

```
    CALCULATE NPERIODS = SUM(SHIFT(Exceed) .GT. Exceed) + \
    Exceed$[Lastday]
ENDIF
IF 'yes' .IN. DURATION
    CALCULATE DPERIODS = SUM(Exceed)
ENDIF
IF 'Tempdiff' .IN. PRINT
    PRINT [IPRINT = *] \
        'Cumulative mean temperature differences from', \
        CRITT,'degrees'
    & ' Day Mean'
    & [IPRINT = *] Day,Tempdiff; FIELDWIDTH = 4,5
ENDIF
IF 'number' .IN. PRINT
    PRINT [IPRINT = *] \
        'Number of periods above critical temperature:',NPERIODS
ENDIF
IF 'duration' .IN. PRINT
    PRINT [IPRINT = *] \
        'Duration of periods above critical temperature:',DPERIODS
ENDIF
ENDPROCEDURE
```

Write the PROCEDURE, OPTION, and PARAMETER statements for this procedure, and the statements to assign identifiers to any parameters that may be unset. Write statements to test the procedure:

(a) with negative differences counted as zero and with the summary values calculated and printed but not returned to the main program, and
(b) with negative differences counted as their actual values and with only the mean temperature differences calculated, and not printed by the procedure but returned to the main program to be printed.

10 Models for time series

10.1 Introduction

A time series is a set of measurements of a variable recorded at successive points in time. This chapter describes a particular class of models for time series called Autoregressive Integrated Moving-Average, or "ARIMA" models. They are designed for the analysis of series of observations taken at regular intervals, such as hourly or yearly, and can describe the behaviour of a single series or relate one series to others.

The theory for these models can be found in many books on the analysis of time series. The standard text is *Time series analysis, forecasting and control* (Box and Jenkins 1970), and a simpler introduction is *Time series analysis and forecasting* (Anderson 1976). The theory is also described, with particular reference to the Genstat facilities in Release 4.04, in *Time series in Genstat* (Tunnicliffe Wilson 1982).

There are many other approaches to the analysis of time series. The simpler methods can be carried out with Genstat by using the CALCULATE directive, which provides several standard functions for convenient manipulation of time series. For an example, see Chapter 5 of *Genstat 5: an introduction*, where a harmonic analysis is performed. The more complex techniques of spectral analysis can also be carried out with Genstat, and these are illustrated in Chapter 11.

10.2 Choosing models with the help of correlograms

In most of the United States of America, telephone users do not pay directly for local calls, but are charged a flat rate every month. One of the major US telephone companies, the General Telephone Electronics Corporation (GTE), started an experiment in 1975 to study the effect of changing this system to include a charge according to the number of calls made. Details of the experiment are given by Jenkins and Wilkinson (1982), and the data were kindly supplied to us by Gwilym Jenkins and Partners, Lancaster. Three telephone exchanges in Illinois were chosen, and calls from domestic telephones were counted from May 1975 until November 1979. During this period, the charging system was changed: in September 1977, a charge per call was introduced at the same time as reducing the "flat rate" charge, and in June 1979 the variable charge was increased by a small amount. The following Genstat output shows how the average number of calls each month per domestic telephone, from May 1975 to November 1979, can be read and displayed.

```
 1  VARIATE [VALUES=1...55] Month
 2  UNITS Month
 3  READ [PRINT=data] Call

 4                                93.3  85.8  89.2  91.5  84.4  89.2  90.9  99.0
 5  100.4  91.2  96.0  92.4  91.6  88.8  86.6  86.7  81.6  88.9  80.6 102.5
 6  122.1  90.7  96.2  91.7  94.1  90.5  90.5  90.4  68.2  76.3  71.9  98.5
 7   87.7  75.8  98.8  82.1  83.5  80.0  79.1  80.4  73.1  78.3  77.0  89.1
 8  103.5  79.4  85.5  83.0  83.0  77.9  77.4  75.6  66.4  76.8  75.3 :
 9  OPEN 'Phone.grd'; CHANNEL=1; FILETYPE=graphics
10  PEN 1; METHOD=line; SYMBOLS=1
11  DGRAPH [TITLE='Number of calls May 1975 to November 1979'] \
12     Call; Month; PEN=1
```

The PEN and DGRAPH statements in this program produce a high-resolution graph, shown in Figure 10.1. The production of high-resolution graphs with Genstat is described in Chapter 3 of *Genstat 5: an introduction*.

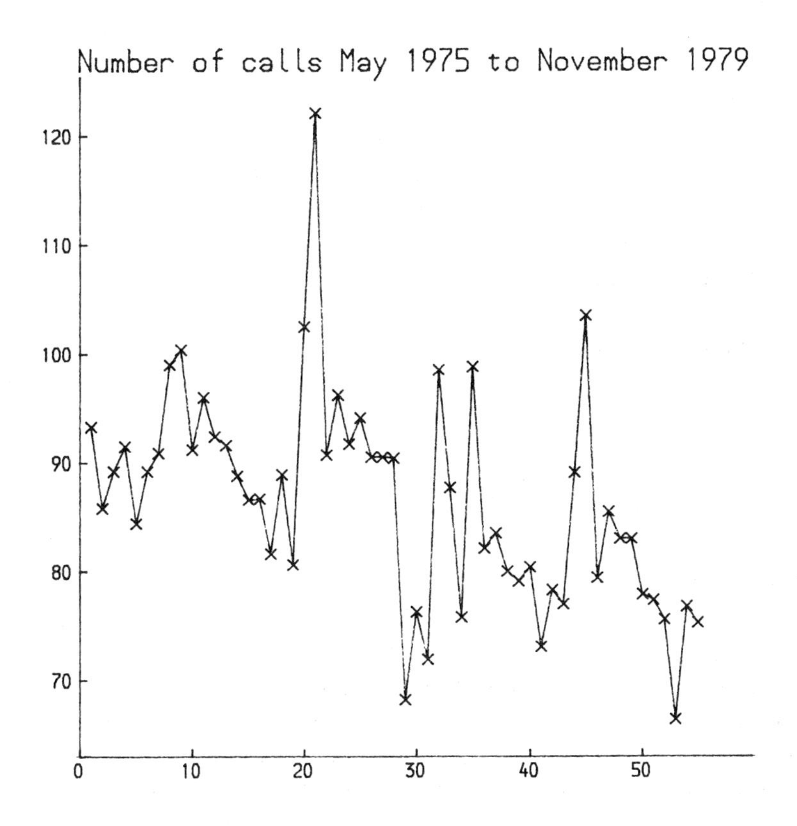

Figure 10.1. High-resolution graph of number of telephone calls.

There are several characteristics of this series. First, the number of local calls dropped as soon as the new charging system was introduced in month 29, and then remained low. Second, there is a regular pattern of variation in the number of calls: each year the number of calls decreased from month to month except during the autumn and early winter when more calls were made. The existence of a regular seasonal pattern is common to many series of this nature, and must be accounted for by including seasonal effects in any model of the series. Thirdly, there were some months, particularly January 1977, when an unusually large number of calls were made. These were probably caused by bad weather preventing people from travelling. Certainly in January 1977 there were record snowfalls, and in March 1978 an ice storm severely curtailed travel in the area.

Before trying to quantify the effects of the changes in the charging system, a reasonable model for the normal month-to-month variation must be chosen. For this purpose, only the data up to September 1977 are relevant: this subset can be identified by the RESTRICT directive. All time-series directives in Genstat allow restrictions, provided the set of units is contiguous: that is, the restriction defines one run of successive dates. Furthermore, the extreme effect of the severe weather in January 1977 should be avoided for the time being. We set the value for this month, the 21st in the series, to be missing, using a CALCULATE statement with a *qualified identifier* that uses the dollar sign to denote the position of a value in a structure: qualified identifiers are described in Chapter 1, Section 1.2. But first we take a copy of the series,because the full set of values will be needed later.

```
CALCULATE Callcopy = Call
CALCULATE Call$[21] = !(*)
RESTRICT Call; Month .LE. 28
```

Notice that we cannot put

```
CALCULATE Call$[21] = *
```

because in a CALCULATE statement the asterisk is taken to mean multiplication unless contained in an unnamed variate like !(*).

This is a short series for investigating any kind of pattern, but nonetheless it is possible to get some indication of what a sensible model would be. We shall first transform the series, using logarithms, in the hope that the resulting series will have roughly constant variance—one of the standard assumptions of ARIMA model analysis. Many series of counts exhibit the property of the standard deviation of an observation increasing with the size of the observation, and the log transformation approximately counteracts this tendency. We shall also try various differencing operations on the series to try to make it *stationary*. This means that its behaviour at any time must be able to be described without reference to the actual time; in particular, the mean value of the series must be constant. Again, ARIMA models are relevant only for series that can be made stationary by such operations.

If a series consists of a steady long-term trend, a constant series will result from taking first differences; that is, replacing each month's value with the difference between that value and the one from the previous month. If the series consists of a regular seasonal fluctuation associated with the calendar year, a constant series will result from taking differences with lag 12: that is, differences between each observation and the observation in the same month of the previous year.

There is a simple function in Genstat to form differenced series, as shown in the following statements:

```
CALCULATE Lcall = LOG(Call)
& Dlcall = DIFFERENCE(Lcall)
& Dslcall = DIFFERENCE(Lcall; 12)
& Ddslcall = DIFFERENCE(Dslcall)
```

The series Ddslcall is twice differenced, once at lag 12 and once at lag 1—the lag number refers to how far back in the series the comparison with each observation is made. Hence, any long-term trend and any seasonal fluctuation have both been removed. Notice that you need to specify only one argument of the DIFFERENCE function if you want first differences. All three differencing operations can be done in a single statement:

```
CALCULATE Dlcall,Dslcall,Ddslcall = \
    DIFFERENCE(Lcall,Lcall,Dslcall; 1,12,1)
```

To see whether there is any further pattern in these series, we display their *autocorrelation functions*. The autocorrelation function of a series is the set of correlations of the series with itself lagged by one observation, two observations, and so on. Because the series is short, there is no point in looking for correlations with a lag of more than about twelve months.

```
13  CALCULATE Callcopy = Call
14  CALCULATE Call$[21] = !(*)
15  RESTRICT Call; Month .LE. 28
16  CALCULATE Lcall = LOG(Call)
17  & Dlcall,Dslcall,Ddslcall = DIFFERENCE(Lcall,Lcall,Dslcall; 1,12,1)
18  CORRELATE [GRAPH=auto; MAXLAG=13] Lcall,Dlcall,Dslcall,Ddslcall
```

Lcall

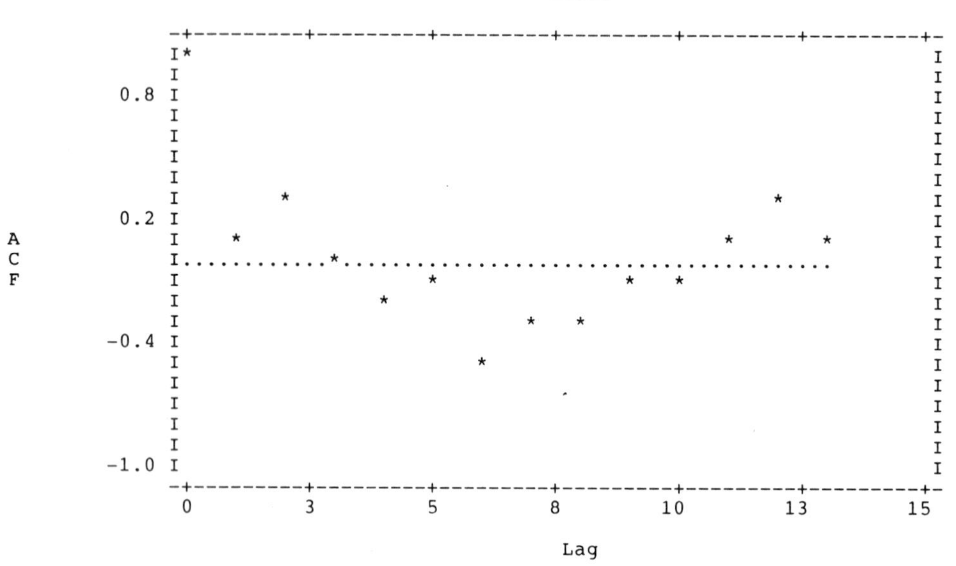
0.8
0.2
-0.4
-1.0
ACF
0
3
5
8
10
13
15
Lag

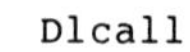
Dlcall

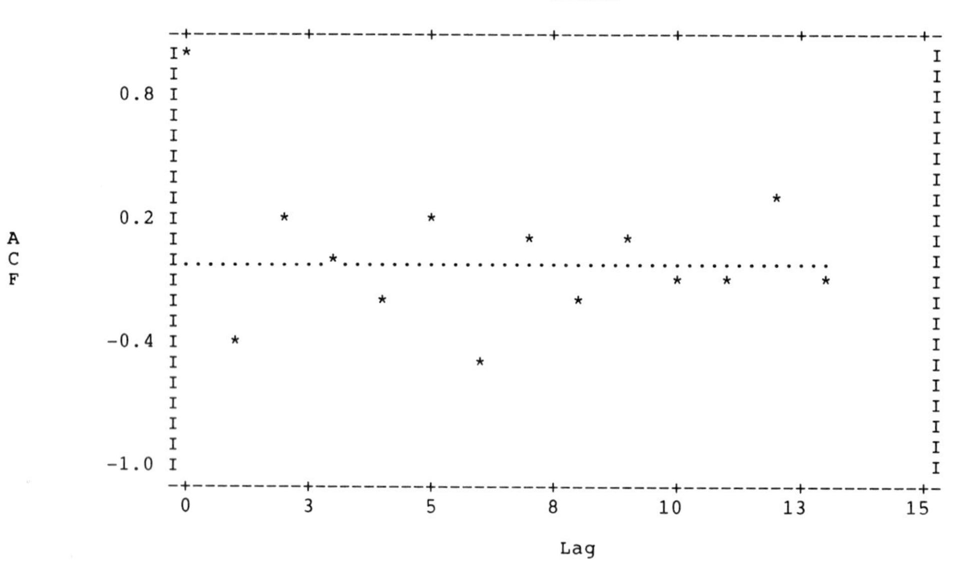
0.8
0.2
-0.4
-1.0
ACF
0
3
5
8
10
13
15
Lag

Dslcall

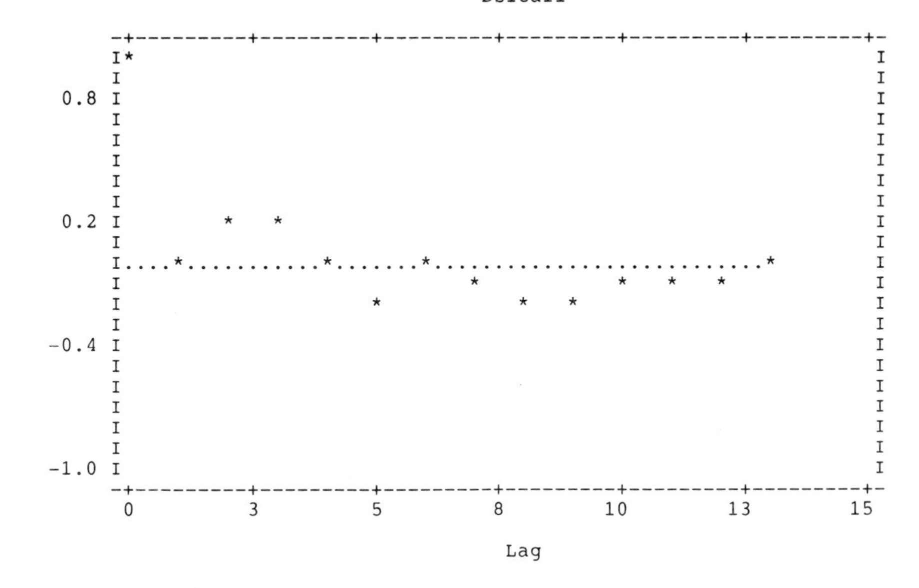

A
C
F

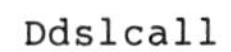

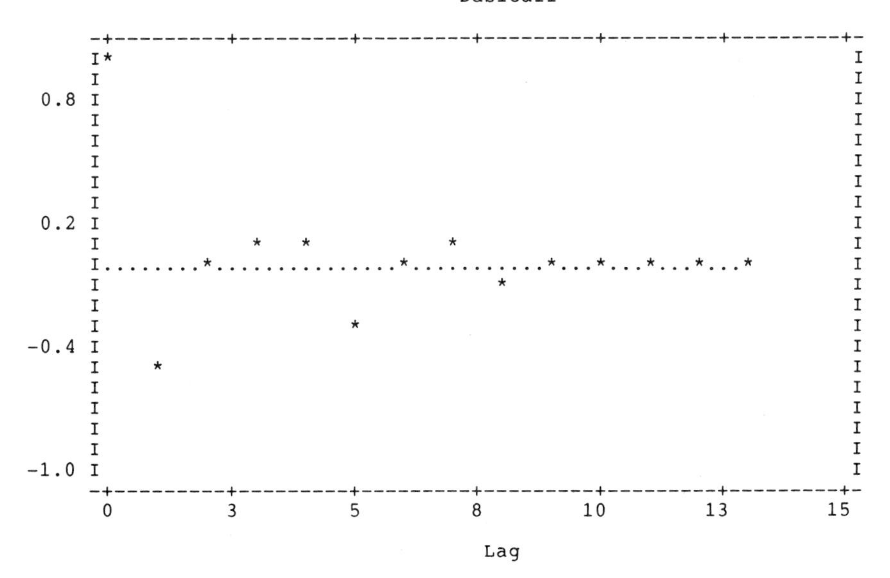

A
C
F

There is no room in this book to discuss the interpretation of correlograms, as these pictures are called. You will find information about them in Box and Jenkins (1978) and Anderson (1976). In general, a strong positive correlation at a particular lag suggests a periodicity of that frequency. It is essential to remember here, though, that the series are very short, and so not much reliance can be placed on any pattern suggested by the graph. Note in particular that the doubly-differenced series has only 14 observations because the differences are not available for the first 13 months.

These correlograms suggest various possible models. The simplest model comes from the correlogram of the seasonally differenced observations, Dslcall: the absence of large correlations suggests that there is no pattern beyond the yearly seasonal effects. Alternatively, the correlogram for the differenced series suggests that there may be correlations at lag 1, 6, and 12. It is not easy to see how to interpret an effect at lag 6 for this series; ignoring this lag would suggest a moving-average model with one non-seasonal term, to account for the correlation at lag 1, and one seasonal term, to account for the correlation at lag 12.

10.3 Estimating parameters in ARIMA models

We shall use Genstat to fit both the models suggested above to the full series, including observations after the price changes. The models clearly will not be able to account for the change, but the analyses will provide necessary information for later comparison with models including an effect of the change.

The first model involves no autocorrelation. We shall further constrain this model by fixing the mean to be zero, because there is no apparent trend with time. If a mean were to be estimated for the differenced series, it would correspond to a constant increase or decrease in the number of calls every year. The fit of this model can therefore be summarized simply by the variance of the differenced series about zero, as calculated by the following Genstat statements.

```
  19   RESTRICT Call
  20   VARIATE Dslcall
  21   CALCULATE Call$[35] = !(*)
  22   & Dslcall = DIFFERENCE(LOG(Call); 12)
  23   & Vardsl = SUM(Dslcall**2)/(55-NMV(Dslcall))
  24   PRINT Vardsl

       Vardsl
     0.006872
```

The standard VARIANCE function is not used here because it calculates variance about the mean rather than about zero.

The model we want to fit is the simplest example of an ARIMA model, requiring no parameters to be estimated, and there is no need to use the special time-series directives to fit it. However, it will serve to introduce ARIMA models in Genstat in a simple context.

There is a special data structure in Genstat provided for the specification of time series models, called a TSM. Here is the declaration of the simplest, null, ARIMA model, which we shall refer to by the identifier Null:

```
VARIATE [VALUES = 0,0,0] Nullord
&       [VALUES = 1,0,*] Nullpar
TSM Null; ORDERS = Nullord; PARAMETERS = Nullpar
```

The three zeros given in the list of orders correspond to the three parts of the term "ARIMA". The first means that there are no autoregressive (AR) parameters in the model, the second that there is no integration (I), and the third that there are no moving-average (MA) parameters. "Integration" corresponds to taking differences of the series: we want none in the model because the series has already been differenced. The meaning of the other types of parameter will be discussed later.

The three parameters are as follows:

1 transformation parameter
0 constant
* variance

The transformation parameter, b say, refers to the Box-Cox family of transformations:

$$T(x) = \begin{cases} (x^b - 1)/b & b > 0 \text{ or } b < 0 \\ \log(x) & b = 0 \end{cases}$$

We have already taken logarithms of the series, so require no further transformation: hence set $b = 1$. The constant is set to 0, and the variance is missing—we shall use a time-series statement to estimate this remaining parameter.

The list of parameters is extended beyond these basic three if either of the AR or MA orders is non-zero, as you will see later.

The orders and parameters can be assigned directly in the TSM statement using the exclamation mark, rather than setting up variates of orders and parameters explicitly as before:

```
TSM Null; ORDERS = !(0,0,0); PARAMETERS = !(1,0,*)
```

The model is fitted to the transformed and differenced series by an ESTIMATE statement. We set the option CONSTANT = fix to fix the constant term at the value already set: 0.0. Here are the TSM and ESTIMATE statements and the resulting output.

```
    25  TSM Null; ORDERS=!(0,0,0); PARAMETERS=!(1,0,*)
    26  ESTIMATE [CONSTANT=fix] Dslcall; TSM=Null


26.........................................................................


******** Warning (Code TS 20). Statement 1 on Line 26
Command: ESTIMATE [CONSTANT=fix] Dslcall; TSM=Null


The iterative estimation process has failed to progress towards a solution
The current cycle number is 1


***** Time-series analysis *****

  Output series: Dslcall
    Noise model: Null
                          autoregressive    differencing  moving-average
          Non-seasonal                 0               0               0


              d.f.      deviance
  Residual     39         0.2680


*** Autoregressive moving-average model ***

Innovation variance 0.006872

                        ref.      estimate          s.e.
Transformation             0       1.00000         FIXED
Constant                   0            0.         FIXED

* Non-seasonal; no differencing
```

A warning is printed here to draw attention to the fact that no iteration is needed to find the solution because there are no autoregressive or moving-average parameters to be estimated. All that has been done is to estimate the variance, 0.006872, which needs no iteration. The other parameters are given numbers for reference: here they are all zero because none are actually estimated.

The deviance shown here is the residual sum of squares, so that the variance is estimated by the deviance divided by the number of degrees of freedom. The residual variance is usually referred to as the *innovation variance* in the analysis of time series, because it is the variance of the residual series, known as the *innovation series*, from whose values the observed series is derived through the model. In this null model, the innovation series is identical to the observed series.

This same model could be fitted directly to the series Call rather than to the differenced logged series:

```
TSM Null2; ORDERS = !(0,0,0, 0,1,0,12); PARAMETERS = !(0,0,*)
ESTIMATE [CONSTANT = fix] Call; TSM = Null2
```

Here the list of orders is extended to include a seasonal component. The extra numbers 0,1,0,12 refer to seasonal orders of type AR, I, and MA at period 12. Hence the model still has no AR or MA parameters, as before, but there is one order of differencing with period 12; that is, at lag 12. Note that the first three orders are always for the non-seasonal part of the model. The parameters are as before, except that the log transformation has been specified by setting the transformation parameter to 0. The model is then fitted to the original series Call rather than to the differenced logged series—hence the result would be exactly as before.

As we said above, using the time-series directives for this first model is like using a sledgehammer to crack a nut. We shall now use the sledgehammer to crack a more complicated time-series problem: the second model mentioned in the last section. This is an ARIMA model for the differenced logged series, involving one moving-average parameter at lag 1 and one at lag 12. A moving-average model seeks to explain the observed series of values as a weighted sum of contributions from the innovation series:

$$y_t = e_t - \theta_1 \times e_{t-1} - \dots - \theta_n \times e_{t-n}$$

Thus the second model has the equation

$$\mathrm{D}(\log(y))_t = e_t - \theta_1 \times e_{t-1} - \theta_{12} \times e_{t-12} + \theta_1 \times \theta_{12} \times e_{t-13}$$

where $\mathrm{D}(\log(y))_t$ stands for the differenced logged series.

By contrast, an autoregressive model fits the observed values to a weighted sum of contributions from previous observations in the same series, together with the current innovation:

$$y_t = \phi_1 \times y_{t-1} + \dots + \phi_n \times y_{t-n} + e_t$$

In fitting these models, the values of θ_i or ϕ_i that minimize the innovation variance are estimated. Many of these will be zero when there are seasonal terms.

The orders for the second model are specified as before:

```
VARIATE [VALUES = 0,1,1, 0,0,1,12] Ord2
```

There are now five parameters, namely the transformation parameter 0, the constant 0, the variance, and the two MA parameters, θ_1 and θ_{12}, that will be fitted by the ESTIMATE statement.

```
27  TSM Null2; ORDERS=!(0,0,0, 0,1,0,12); PARAMETERS=!(0,0,*)
28  VARIATE [VALUES=0,1,1, 0,0,1,12] Ord2
29  & [VALUES=0,0,3(*)] Par2
30  TSM Masma; ORDERS=Ord2; PARAMETERS=Par2
31  ESTIMATE [CONSTANT=fix] Call; TSM=Masma
```

```
31.......................................................................

***** Time-series analysis *****

  Output series: Call
    Noise model: Masma
                          autoregressive    differencing  moving-average
          Non-seasonal                 0               1               1
          Period 12                    0               0               1


              d.f.      deviance
  Residual     50         0.3180


*** Autoregressive moving-average model ***

Innovation variance 0.005728

                    ref.      estimate          s.e.
Transformation        0             0.          FIXED
Constant              0             0.          FIXED

* Non-seasonal; differencing order 1

                lag ref.      estimate          s.e.
Moving-average    1    1         0.571          0.122

* Seasonal; period 12; no differencing

                lag ref.      estimate          s.e.
Moving-average   12    2        -0.597          0.189
```

This model appears to fit slightly better than the first, because the variance is smaller: 0.0057 compared to 0.0069. Also, the two moving-average parameters are much larger than their standard errors. If all we were interested in was an ARIMA model for the series, we should go further and look at the residual series for any remaining pattern, part of the model-checking process described in Box and Jenkins (1970). However, here we know that this description of the series is unlikely to be sufficient because of the pronounced change in the series due to the price change, so we shall leave the investigation of residuals until we have fitted a more satisfactory model.

10.4 Intervention analysis

The effect of the price changes can be estimated by incorporating *intervention variables* in the model for the number of calls. For example, suppose that the effect of the price change is to decrease the number of calls by a certain amount—constant up to the next price change. This simple effect is included by putting into the model a contribution from a step variable that is 0 up to August 1977 and 1 from September

1977 onwards. Similarly for the effect of the price change in June 1979. Therefore, we shall start by considering the model

$$\log(\text{Calls}) = a \times \text{Step1} + b \times \text{Step2} + \text{Noise}$$

where Step1 and Step2 are the two intervention variables, a and b are parameters to be estimated, and Noise is one of the models for the number of calls investigated in the last section (so called because it tends to drown out the information for which the telephone company is "listening").

The effects of the intervention variables are specified in Genstat by the directive TRANSFERFUNCTION (which we abbreviate to TRANSFER):

```
VARIATE Change[1,2]; VALUES = !(28(0),27(1)),!(49(0),6(1)); \
    DECIMALS = 0
TRANSFER Change[1,2]
```

The model for each intervention variable shown above is a particularly simple case of a *transfer-function model* relating an input series to the output series: the output series here being the numbers of calls. You will see later how more general transfer-function models can be specified. After a TRANSFER statement, an ESTIMATE statement will include all the effects when fitting the noise model; here we use the null ARIMA model, Null2 defined above, incorporating seasonal differencing.

```
  32  VARIATE Change[1,2]; VALUES=!(28(0),27(1)),!(49(0),6(1)); \
  33     DECIMALS=0
  34  TRANSFER Change[1,2]
  35  ESTIMATE [CONSTANT=fix] Call; TSM=Null2


35.............................................................................


***** Time-series analysis *****

  Output series: Call
    Noise model: Null2
                               autoregressive     differencing  moving-average
            Non-seasonal                    0                0               0
            Period 12                       0                1               0
     Input series:
Change[1]        ; transfer function:
            Non-seasonal                    0                0               0
Change[2]        ; transfer function:
            Non-seasonal                    0                0               0


                d.f.      deviance
  Residual        39       0.09632


*** Transfer-function model 1 ***

Delay time 0

                        ref.      estimate          s.e.
Transformation             0       1.00000         FIXED
Constant                   0            0.         FIXED
```

```
* Non-seasonal; no differencing

                 lag ref.      estimate          s.e.
Moving-average    0    1       -0.1271         0.0150

*** Transfer-function model 2 ***

Delay time 0

                     ref.      estimate          s.e.
Transformation         0       1.00000          FIXED
Constant               0            0.          FIXED

* Non-seasonal; no differencing

                 lag ref.      estimate          s.e.
Moving-average    0    2       -0.0413         0.0203

*** Autoregressive moving-average model ***

Innovation variance 0.002470

                     ref.      estimate          s.e.
Transformation         0            0.          FIXED
Constant               0            0.          FIXED

* Non-seasonal; no differencing

* Seasonal; period 12; differencing order 1
```

Notice that the residual variance for this model is 0.0025. This is a distinct improvement on 0.0069, estimated for this model with no allowance for the price changes. The coefficients of the step variables are shown on the output as moving-average parameters at lag 0: the number of calls in a month is related to the value of the step variables in that month. Comparison with their standard errors shows that, with this model for the noise, there is a very significant drop in the number of calls per month after the first change. The size of the drop on the natural scale is found by taking the antilog of the coefficient (-0.127): the number of calls reduces by a factor of 0.88; that is, it reduces by 12%. There is rather inconclusive evidence of a drop of about 4% after the second price change.

However, all this interpretation depends on the assumption of the noise model. When fitting any kind of statistical model, you should always try to check the adequacy of the model as far as you can. One way is to consider alternative models that extend the fitted model in some way, and see whether there is evidence for the extension being justified. We could fit the second model considered in the last section, or several other extensions that involve adding one or two parameters to describe autocorrelation in the series. We shall consider one extension in particular: a model with just one seasonal moving-average term. This model is declared by the following statement:

```
TSM Sma; ORDERS = !(0,0,0, 0,1,1,12); PARAMETERS = !(0,0,*,*)
```

While fitting this model, we shall include two further intervention variables in order to try to quantify the effect of the two storms whose effect we have previously eliminated. These can be represented by *spike* variables as follows:

```
VARIATE Snow,Ice; VALUES = !(20(0),1,34(0)),!(34(0),1,20(0)); \
    DECIMALS = 0
```

The inclusion of these variables, together with the observations in January 1977 and March 1978 which we retrieve from the copy of the original series, will not change the fit of the model, since one parameter is estimated for each of the observations. However, the parameter estimates will quantify the effects of the weather.

The results of fitting this model are shown below.

```
  36   TSM Sma; ORDERS=!(0,0,0, 0,1,1,12); PARAMETERS=!(0,0,*,*)
  37   VARIATE Snow,Ice; VALUES=!(20(0),1,34(0)),!(34(0),1,20(0)); \
  38      DECIMALS=0
  39   CALCULATE Call = Callcopy
  40   TRANSFER Change[1,2],Snow,Ice
  41   ESTIMATE [CONSTANT=fix] Call; TSM=Sma


41..............................................................................


***** Time-series analysis *****

  Output series: Call
    Noise model: Sma
                        autoregressive    differencing  moving-average
          Non-seasonal               0               0               0
          Period 12                  0               1               1
     Input series:
Change[1]        ; transfer function:
          Non-seasonal               0               0               0
Change[2]        ; transfer function:
          Non-seasonal               0               0               0
Snow             ; transfer function:
          Non-seasonal               0               0               0
Ice              ; transfer function:
          Non-seasonal               0               0               0


              d.f.      deviance
  Residual      38       0.07922


*** Transfer-function model 1 ***

Delay time 0

                      ref.      estimate         s.e.
Transformation           0       1.00000        FIXED
Constant                 0            0.        FIXED

* Non-seasonal; no differencing

                  lag ref.      estimate         s.e.
Moving-average     0     1       -0.1179       0.0113
```

```
*** Transfer-function model 2 ***

Delay time 0

                        ref.      estimate          s.e.
Transformation            0        1.00000          FIXED
Constant                  0             0.          FIXED

* Non-seasonal; no differencing

                    lag ref.      estimate          s.e.
Moving-average       0    2        -0.0312         0.0183


*** Transfer-function model 3 ***

Delay time 0

                        ref.      estimate          s.e.
Transformation            0        1.00000          FIXED
Constant                  0             0.          FIXED

* Non-seasonal; no differencing

                    lag ref.      estimate          s.e.
Moving-average       0    3         0.1525         0.0433


*** Transfer-function model 4 ***

Delay time 0

                        ref.      estimate          s.e.
Transformation            0        1.00000          FIXED
Constant                  0             0.          FIXED

* Non-seasonal; no differencing

                    lag ref.      estimate          s.e.
Moving-average       0    4         0.1452         0.0433


*** Autoregressive moving-average model ***

Innovation variance 0.001451

                        ref.      estimate          s.e.
Transformation            0             0.          FIXED
Constant                  0             0.          FIXED

* Non-seasonal; no differencing

* Seasonal; period 12; differencing order 1

                    lag ref.      estimate          s.e.
Moving-average      12    5          0.937          0.245
```

The residual variance is now 0.0015, so this model seems to fit better than the last. The estimates of the effects of the price changes have changed a little, from 12% to 11% for the first change, and from 4% to 3% for the second change. The effects of the two periods of bad weather seem to be about the same, causing a 16% increase in the number of calls.

However, there is more to accepting a model than just looking at the residual variance. An important check to make is to look for any pattern remaining in the residual series, which can be done by treating this series in the same way as when models were being chosen for the original series. The residual series can be accessed either by setting the RESIDUALS parameter of the ESTIMATE statement, or of a subsequent TKEEP statement. The TKEEP directive accesses information about the latest model fitted by an ESTIMATE statement.

```
42  TKEEP RESIDUALS=Res
43  FACTOR [LEVELS=!(1975...1979); VALUES=8(1975),12(1976...1978), \
44     11(1979)] Year; DECIMALS=0
45  FACTOR [LABELS=!T(Jan,Feb,Mar,Apr,May,Jun,Jul,Aug,Sep,Oct,Nov,Dec); \
46     VALUES=5...12,(1...12)3,1...11] Mon
47  PRINT Mon,Year,Call,Change[],Snow,Ice,Res; FIELDWIDTH=5,5,6(10)

Mon Year       Call Change[1] Change[2]      Snow       Ice       Res
May 1975      93.30         0         0         0         0   0.00047
Jun 1975      85.80         0         0         0         0  -0.03733
Jul 1975      89.20         0         0         0         0   0.00213
Aug 1975      91.50         0         0         0         0   0.02376
Sep 1975      84.40         0         0         0         0   0.04874
Oct 1975      89.20         0         0         0         0   0.01090
Nov 1975      90.90         0         0         0         0   0.06469
Dec 1975      99.00         0         0         0         0  -0.03982
Jan 1976     100.40         0         0         0         0  -0.04320
Feb 1976      91.20         0         0         0         0   0.02316
Mar 1976      96.00         0         0         0         0  -0.00137
Apr 1976      92.40         0         0         0         0  -0.00064
May 1976      91.60         0         0         0         0  -0.01795
Jun 1976      88.80         0         0         0         0  -0.00061
Jul 1976      86.60         0         0         0         0  -0.02759
Aug 1976      86.70         0         0         0         0  -0.03161
Sep 1976      81.60         0         0         0         0   0.01194
Oct 1976      88.90         0         0         0         0   0.00684
Nov 1976      80.60         0         0         0         0  -0.05964
Dec 1976     102.50         0         0         0         0  -0.00258
Jan 1977     122.10         0         0         1         0   0.00274
Feb 1977      90.70         0         0         0         0   0.01621
Mar 1977      96.20         0         0         0         0   0.00080
Apr 1977      91.70         0         0         0         0  -0.00820
May 1977      94.10         0         0         0         0   0.01011
Jun 1977      90.50         0         0         0         0   0.01839
Jul 1977      90.50         0         0         0         0   0.01820
Aug 1977      90.40         0         0         0         0   0.01216
Sep 1977      68.20         1         0         0         0  -0.05032
Oct 1977      76.30         1         0         0         0  -0.02855
Nov 1977      71.90         1         0         0         0  -0.05224
Dec 1977      98.50         1         0         0         0   0.07565
Jan 1978      87.70         1         0         0         0  -0.05802
Feb 1978      75.80         1         0         0         0  -0.04640
Mar 1978      98.80         1         0         0         1   0.00005
Apr 1978      82.10         1         0         0         0  -0.00040
May 1978      83.50         1         0         0         0   0.00783
Jun 1978      80.00         1         0         0         0   0.01178
Jul 1978      79.10         1         0         0         0   0.00029
Aug 1978      80.40         1         0         0         0   0.01204
Sep 1978      73.10         1         0         0         0   0.02223
Oct 1978      78.30         1         0         0         0  -0.00088
Nov 1978      77.00         1         0         0         0   0.01957
Dec 1978      89.10         1         0         0         0  -0.02940
Jan 1979     103.50         1         0         0         0   0.11127
Feb 1979      79.40         1         0         0         0   0.00292
Mar 1979      85.50         1         0         0         0   0.00071
Apr 1979      83.00         1         0         0         0   0.01053
May 1979      83.00         1         0         0         0   0.00133
Jun 1979      77.90         1         1         0         0   0.01562
```

```
Jul 1979     77.40     1     1     0     0   0.00973
Aug 1979     75.60     1     1     0     0  -0.01909
Sep 1979     66.40     1     1     0     0  -0.04412
Oct 1979     76.80     1     1     0     0   0.01101
Nov 1979     75.30     1     1     0     0   0.02719

48  CORRELATE [GRAPH=auto; MAXLAG=25] Res; TEST=Chi
```

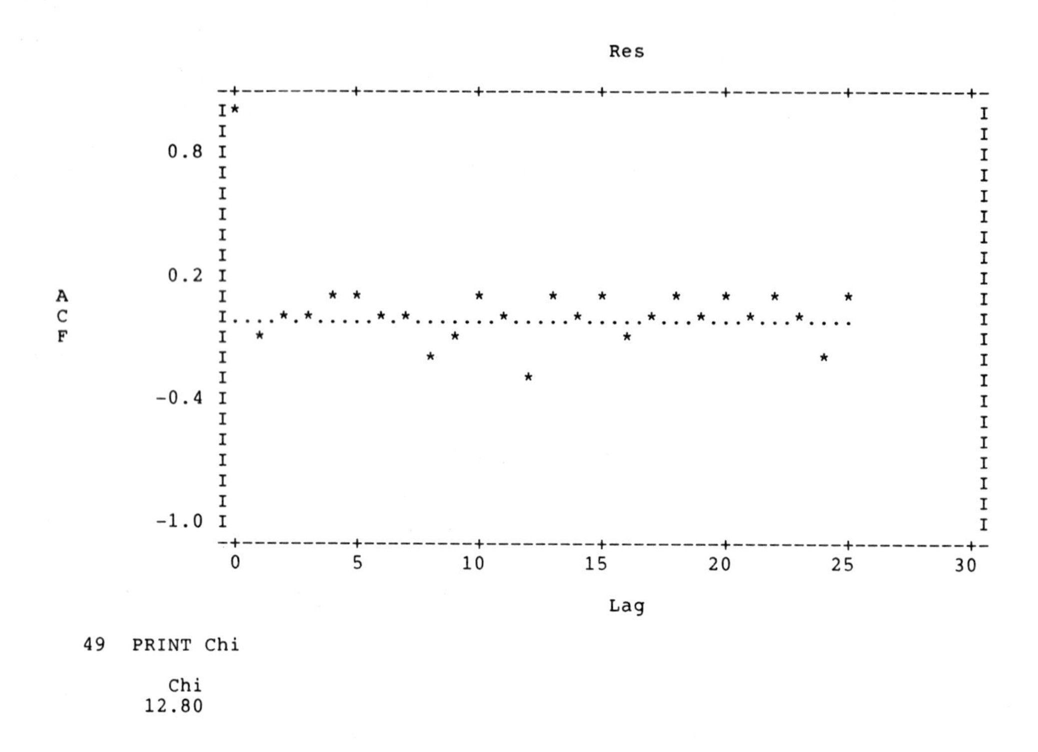


```
49  PRINT Chi

      Chi
    12.80
```

The graph shows no systematic pattern of autocorrelation in the residual series. This is confirmed by the value of the statistic stored in the scalar Chi by the TEST parameter of the CORRELATE statement: if there is no autocorrelation up to the maximum lag, the statistic has an approximate chi-squared distribution with degrees of freedom equal to the maximum lag.

The listing of the residuals shows that one observation is not well fitted by the model: the number of calls in January 1979 is about 12% larger than expected under the model (1.12 is the antilog of the residual 0.111). This discrepancy could well be due to more bad weather, which was not bad enough to get into the records.

Since the last extension to the model improved the fit, it is worth considering further extensions. One possibility is to fit a more complicated transfer-function model for the relationship between the number of calls and the price change. We shall allow the effect of the change to vary from month to month after the change, according to a simple moving-average model (closely analogous to the ARIMA noise model for the number of calls). The model is represented by the equation:

$$\log(\text{Calls})_t = a_1 \times \text{Step } 1_t + a_2 \times \text{Step } 1_{t-1} + b \times \text{Step } 2_t$$

The model thus corresponds to assuming that there is an immediate effect of the first price change and a modified effect after the first month, but the number of calls then stabilizes. The effect of the second price change is taken to be immediate.

The simplest type of transfer-function model, used in the previous model, does not have to be explicitly declared. More complicated transfer-function models are declared in the same way as ARIMA models for the structure of one series, as follows:

```
TSM [MODEL = transfer] Tranma; ORDERS = !(0,0,0,1); \
    PARAMETERS = !(1,0,*,*)
```

The MODEL option must be set to specify that this is a transfer-function model; the orders are then interpreted as follows:

0 there is no delay in the effect of the intervention;
0 there is no autoregressive pattern in the variation of the effect of the intervention;
0 there is no differencing of the intervention series in the model for its relationship with the number of calls;
1 there is one moving-average parameter (hence at lag 1) in the model describing the variation of the effect.

The parameters too are slightly different from ARIMA parameters. There is no variance parameter, since the intervention series are treated as fixed observations, but the other parameters correspond to ARIMA parameters:

1 there is to be no transformation of the intervention series in the model;
0 there is no additive constant;
* this parameter corresponds to lag 0; that is, it is the coefficient in the equation relating the number of calls to the value of the step variable in the current month; it is to be fitted by ESTIMATE;
* this is the MA parameter at lag 1, representing an extra effect of the previous month's value of the step variable; that is, it corresponds to a lagged effect of the price change; it too is to be fitted.

The fit of this model is shown below.

```
50  TSM [MODEL=transfer] Tranma; ORDERS=!(0,0,0,1); \
51      PARAMETERS=!(1,0,*,*)
52  TRANSFER Change[],Snow,Ice; TRANSFER=Tranma,*,*,*
53  ESTIMATE [CONSTANT=fix] Call; TSM=Sma
```

```
53.......................................................................

***** Time-series analysis *****

  Output series: Call
    Noise model: Sma
                        autoregressive    differencing  moving-average
          Non-seasonal               0               0               0
          Period 12                  0               1               1
   Input series:
Change[1]      ; transfer function: Tranma
          Non-seasonal               0               0               1
Change[2]      ; transfer function:
          Non-seasonal               0               0               0
Snow           ; transfer function:
          Non-seasonal               0               0               0
Ice            ; transfer function:
          Non-seasonal               0               0               0


              d.f.     deviance
  Residual      37      0.07498


*** Transfer-function model 1 ***

Delay time 0

                    ref.     estimate          s.e.
Transformation         0      1.00000         FIXED
Constant               0           0.         FIXED

* Non-seasonal; no differencing

                lag ref.     estimate          s.e.
Moving-average    0    1      -0.1750        0.0411
                  1    2      -0.0600        0.0417


*** Transfer-function model 2 ***

Delay time 0

                    ref.     estimate          s.e.
Transformation         0      1.00000         FIXED
Constant               0           0.         FIXED

* Non-seasonal; no differencing

                lag ref.     estimate          s.e.
Moving-average    0    3      -0.0355        0.0183

*** Transfer-function model 3 ***


Delay time 0

                    ref.     estimate          s.e.
Transformation         0      1.00000         FIXED
Constant               0           0.         FIXED

* Non-seasonal; no differencing

                lag ref.     estimate          s.e.
Moving-average    0    4       0.1546        0.0427
```

```
*** Transfer-function model 4 ***

Delay time 0

                      ref.      estimate          s.e.
Transformation           0       1.00000         FIXED
Constant                 0            0.         FIXED

* Non-seasonal; no differencing

                  lag ref.      estimate          s.e.
Moving-average      0    5        0.1433        0.0427


*** Autoregressive moving-average model ***

Innovation variance 0.001433

                      ref.      estimate          s.e.
Transformation           0            0.         FIXED
Constant                 0            0.         FIXED

* Non-seasonal; no differencing

* Seasonal; period 12; differencing order 1

                  lag ref.      estimate          s.e.
Moving-average     12    6         0.920         0.247
```

The residual variance hardly changes at all, from 0.00145 to 0.00143, and the standard error of the extra moving-average type parameter in the first transfer-function model (0.042) is nearly as large as the absolute value of the estimate (0.060). Hence we conclude that this particular extension of the last model is not worthwhile.

10.5 Forecasting

The models above were fitted for the purpose of quantifying the effect of price changes to help decide whether to make similar changes in other telephone areas. Often in the analysis of a time series, however, the goal of an analysis is a forecast of the future behaviour of the series. We shall now use the model that was finally chosen—the model with one seasonal moving-average parameter and four intervention variables with constant effects—to show how to produce forecasts from a model.

There is a directive in Genstat called FORECAST which is designed to form forecasts from any model fitted by the ESTIMATE directive. By default, any FORECAST statement will operate with the model fitted by the latest ESTIMATE statement in the program. If you look back to the previous section, you will see that the last model fitted was the extended model in which the effect of the first price change was allowed to vary with a simple moving-average behaviour. The model we want to base forecasts on is the one fitted previously, using the statements in lines 40 and 41:

```
TRANSFER Change[1,2],Snow,Ice
ESTIMATE [CONSTANT = fix] Call; TSM = Sma
```

This ESTIMATE statement could be given again, followed by a FORECAST statement. Alternatively, if it was guessed in advance that this model was to be chosen for forecasting, an option of that ESTIMATE statement could have been set to save the necessary information about that model

```
ESTIMATE [CONSTANT = fix; SAVE = Finalmod] Call; TSM = Sma
```

The model information could then be accessed using the equivalent SAVE option in FORECAST. There are also SAVE options in TDISPLAY and TKEEP, so that you can display or save results from other models besides that which was last fitted.

The model is based on four intervention variables, so to produce a forecast we must provide future values for them. We shall look 12 months ahead, on the assumption that there are no further price changes or extremes of weather. The following statements give the forecasts shown below, which are also displayed in graphical form.

```
  54  TRANSFER Change[1,2],Snow,Ice
  55  ESTIMATE [PRINT=*; CONSTANT=fix] Call; TSM=Sma
  56  VARIATE [VALUES=12(1)] Nchange[1,2]
  57  & [VALUES=12(0)] Nsnow,Nice
  58  FORECAST [MAXLEAD=12; FORECAST=Ncall; LOWER=Low90%; UPPER=Up90%] \
  59     Nchange[ ],Nsnow,Nice

59..............................................................


*** Forecasts ***

Maximum lead time: 12

    Lead time      forecast  lower limit  upper limit
            1         88.78        83.41        94.49
            2         90.38        84.91        96.19
            3         76.76        72.11        81.69
            4         82.82        77.81        88.15
            5         79.66        74.85        84.79
            6         80.35        75.49        85.52
            7         76.76        72.12        81.70
            8         76.70        72.06        81.63
            9         76.97        72.31        81.92
           10         69.21        65.03        73.66
           11         76.01        71.42        80.90
           12         73.40        68.97        78.13

  60  VARIATE [VALUES=1...12] Nextyear
  61  PEN 1,2; METHOD=line,point; SYMBOLS=0,1; COLOUR=1
  62  DGRAPH [TITLE='Forecast for December 1979 to November 1980'] \
  63     Up90%,Ncall,Low90%; Nextyear; PEN=1,2,1
  64  STOP
```

The graph is shown in Figure 10.2. It includes upper and lower 90% confidence intervals to show the precision of the estimates of future phone use based on the available data.

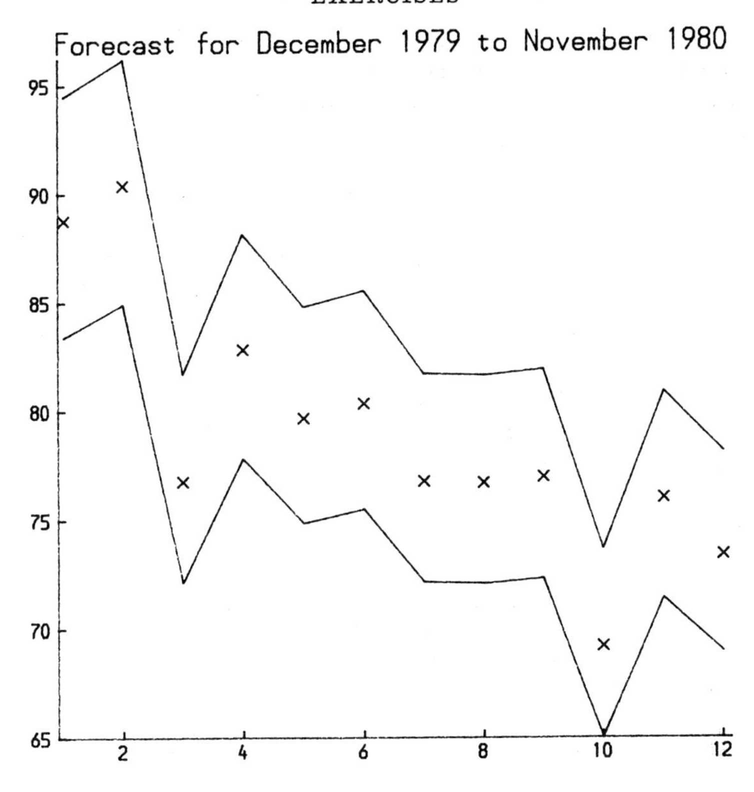

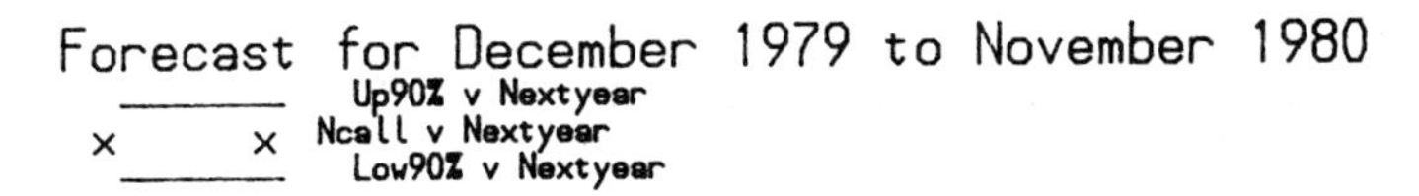

Figure 10.2. High-resolution graph of forecasts of number of telephone calls.

10.6 Exercises

10(1) Generate a random series with 1000 values using the Genstat pseudo-random number generator: give the statement

```
HELP systemfunction,URAND
```

in a Genstat program to find out how to use the generator. Check the adequacy of the generator, which should produce an uncorrelated series, by displaying the autocorrelation function of the random series. Fit an ARIMA model with just one autoregressive parameter to the series, and compare the estimate of the parameter with its standard error.

10(2) The following temperatures (°F) were recorded each day at noon on Ben Nevis, Scotland, between February 1st and June 21st, 1884.

22.9	13.8	31.4	30.1	31.8	31.0	18.3	23.8	31.6	18.9		February
18.8	31.8	31.1	27.0	23.0	19.9	21.5	23.8	22.8	24.8		
20.4	25.8	24.3	22.9	23.2	23.8	19.2	18.2	14.5			
22.6	20.1	18.6	24.9	25.2	30.1	29.4	23.3	20.4	17.6		March
14.7	24.5	32.4	32.0	33.4	37.3	36.7	31.3	26.6	23.8		
21.9	26.9	25.0	25.8	30.5	25.1	21.7	20.7	21.2	22.4	24.2	
25.8	29.1	35.6	32.2	29.9	30.5	32.6	29.0	33.2	35.4		April
32.8	27.4	25.3	27.6	29.3	26.7	24.5	23.2	30.7	26.5		
23.9	24.6	24.4	28.1	26.0	27.3	27.7	27.7	27.0	24.2		
26.3	25.7	25.3	27.9	27.5	25.0	23.6	31.6	35.7	35.2		May
38.8	31.2	32.5	32.0	33.1	31.8	35.0	27.7	31.1	26.7		
30.9	38.7	41.8	45.6	36.0	38.5	46.1	46.0	42.5	35.3	36.8	
33.2	34.5	39.5	38.7	32.5	33.0	33.0	30.9	30.7	37.0		June
36.8	34.7	35.7	31.7	32.0	37.9	42.4	38.2	35.3	38.4	40.9	

(Data from Buchan, 1890.)

Plot the series, the first differences of the series, and the second differences. Choose the least amount of differencing that appears to remove the long-term trend in the temperatures. With the chosen degree of differencing, fit moving-average models with from one up to three parameters. Choose the most suitable model and examine the residuals.

10(3) The temperatures on Ben Nevis between 22nd June and 18th August 1884 were as follows:

39.1	34.1	41.9	32.1	41.9	45.7	56.5	49.4	46.3			June
50.9	47.1	50.9	53.9	50.9	45.9	43.1	45.0	48.3	48.6		July
40.9	41.6	45.7	40.6	40.6	37.0	36.1	33.9	32.1	34.1		
34.2	41.9	39.8	35.1	31.9	34.9	39.8	39.0	42.9	44.4	42.7	
45.2	41.0	37.0	37.5	41.0	45.0	47.9	56.5	55.6	54.2		August
52.5	52.4	44.9	40.9	41.1	44.8	46.7	49.4				

Fit the model, as chosen above, to the whole series, and check how much the parameter estimates change with the additions of the later observations. Include the effect of a step variable in the model to see if there is a difference between the period in which day-length is increasing (up to 21st June) and that in which day-length is decreasing.

11 Spectral analysis of time series

11.1 Introduction

The use of ARIMA models, described in Chapter 10, is only one of many ways to describe patterns in time series. That approach is concerned with modelling the behaviour of a series in the time domain: the behaviour of an observation is described in terms of its relationship with observations that precede it in time. An alternative approach, called spectral analysis, is concerned with modelling behaviour in the frequency domain: the behaviour of the series as a whole is described in terms of effects associated with various frequencies in time.

Spectral analysis is closely related to the method of harmonic analysis which is described in Chapter 5 of *Genstat 5: an introduction*. There, a series of observations is modelled by a sum of a few sinusoidal curves with a common base frequency; in spectral analysis, a whole range of frequencies is employed, possibly giving information about hidden periodicities in the series, and providing a useful description of the series—its spectrum.

For a description of the background and theory of spectral analysis see *The analysis of time series* (Chatfield 1980).

11.2 Looking for cyclical structure in a time series

The Earth is slowing in its rotation about its axis. The length of a day is increasing by about ten microseconds every year, but this increase is not steady. The mean daylengths in each year from 1821 to 1970 are given by Luo *et al.* (1977); the values were derived from various sources, particularly from observations of eclipses before the establishment of the atomic clock, and smoothed to give a series with approximately constant precision of measurement. The following output displays the change in daylength over the 150-year period.

```
  1  "Mean day lengths, 1821 - 1970"
  2  VARIATE [VALUES=1821 ... 1970] Year
  3  UNITS Year
  4  READ [PRINT=data,errors] Lday

  5  -217  -177  -166  -136  -110   -95   -64   -37
  6   -14   -25   -51   -62   -73   -88  -113  -120
  7   -83   -33   -19    21    17    44    44    78
  8    88   122   126   114    85    64    55    51
  9    40    30    14     1     1    -4   -13   -56
```

```
10     -83   -104    -93    -88    -75    -80   -101   -156
11    -226   -293   -333   -347   -329   -279   -205   -131
12     -86    -59    -48    -35    -30    -12     11     57
13      92     86     53     26      6    -12    -35    -31
14       0     36     54     65    104    166    248    318
15     384    415    421    402    392    387    391    396
16     400    391    361    328    296    282    269    256
17     225    202    193    205    201    178    139    130
18     101     67     22      2     12     26     21     10
19     -11    -12    -15      6     22     51     78    111
20     141    150    157    148    138    137    151    151
21     136    111    105    105    110    104     92     96
22     115    144    126    131    112    119    139    183
23     206    231    244    239    263    273 :
24  GRAPH [NROWS=21; NCOLUMNS=61] Lday; Year
```

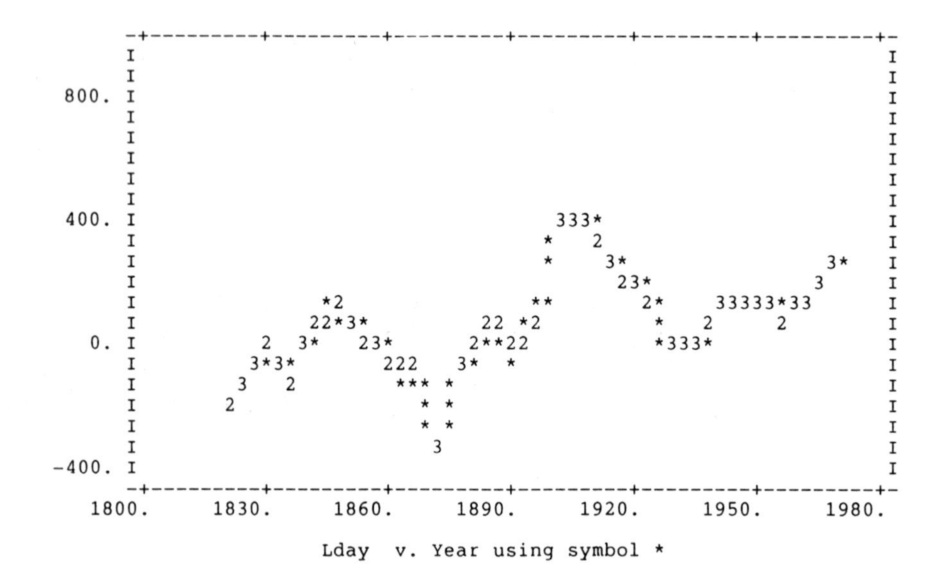

The daylength is given as differences from the mean daylength in 1893, in units of 10 microseconds. The graph shows that there is indeed considerable fluctuation about the increasing trend in the series. There are many possible sources of this variation, for there are several astronomical bodies, particularly the Sun and the Moon, that are likely to exert some effect on the Earth's rotation. Some evidence about which of these are having an important effect can be gained by finding cyclical behaviour in the series at frequencies associated with the behaviour of these other bodies.

The cyclical behaviour of a series with a noticeable trend is likely to be dominated by low frequency terms, because the trend will be modelled by cyclic variations with very long periods. Therefore, it is sensible to remove the trend by differencing the series, as described in Chapter 10, Section 10.2. Cyclic variation with shorter periods will be largely unaffected by this differencing operation.

```
25  CALCULATE Difflday = DIFFERENCE(Lday)
26  GRAPH [NROWS=21; NCOLUMNS=61] Difflday; Year
```

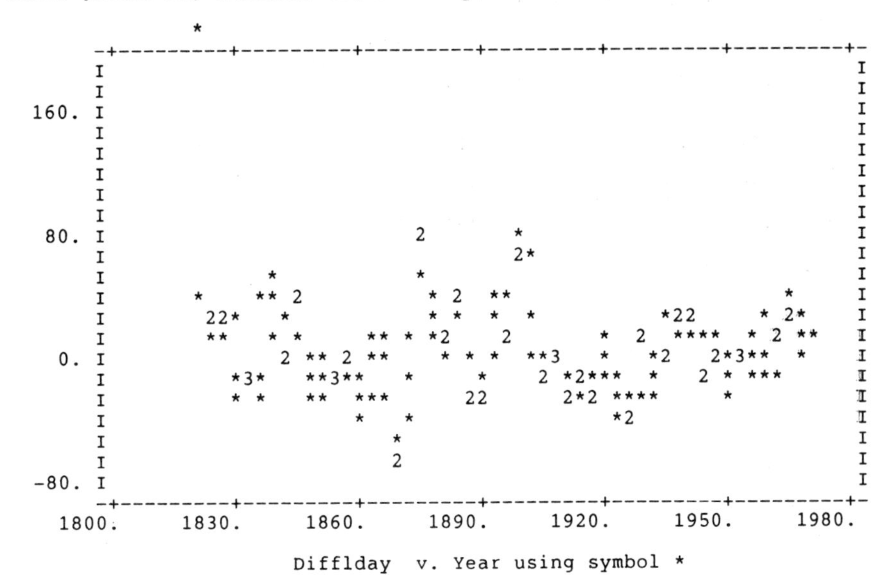

In the last chapter, the behaviour of time series was described using the autocorrelation function. This function is also needed in the calculation of the spectral density function, which is the goal of spectral analysis. The autocorrelations are displayed in a correlogram below, using the CORRELATE directive as in Chapter 10, Section 10.2.

```
27  CORRELATE [MAXLAG=25; GRAPH=autocorrelations] Difflday; \
28      AUTOCORRELATIONS=Auto
```

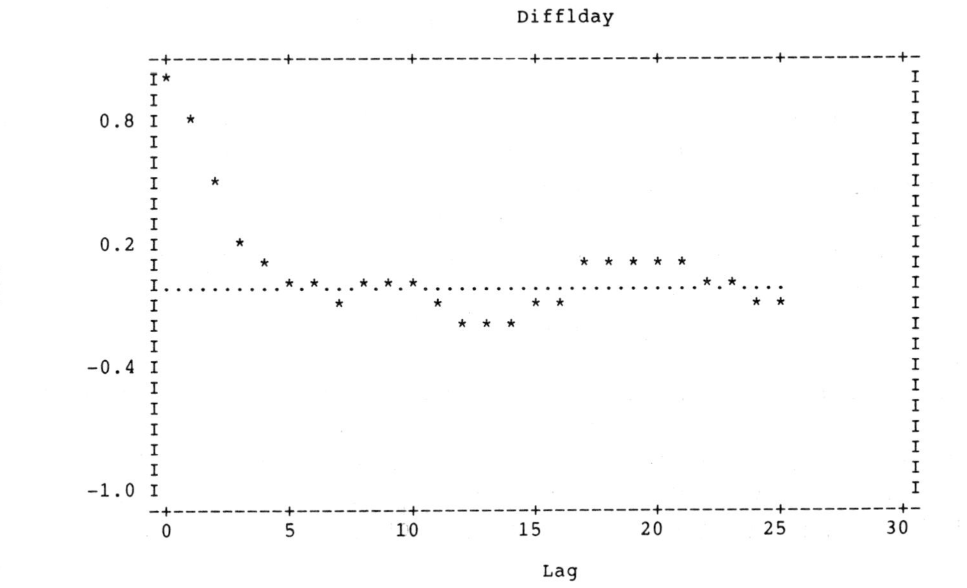

The graph shows that there is considerable autocorrelation between successive values even of the differenced series. There is no evidence of simple cyclic behaviour, which would be marked by relatively large autocorrelations near some lags.

The spectrum of the time series can display more complex behaviour, when there may be several cycles with different periods. The spectrum is more correctly called the power spectral density function; it is a function of frequency and represents the contribution to the variance of the time series of components with varying frequency. The area under the spectral density function is therefore the total variance of the time series. The best known spectrum is that of light; the spectrum of a time series describes the variation in the series in terms of cyclical frequencies in the same way as the spectrum of a light-source describes the make-up of the light in terms of colours. A prism is needed to demonstrate this to the human eye, but the human ear performs a similar analysis of sound automatically, so that a musical chord is perceived as a combination of several tones with different frequencies.

We shall work with the normalized spectrum, under which the area is 1, which describes the proportion of variance attributable to components with varying frequency. The normalized spectrum is the Fourier transform of the autocorrelations: it is defined for a certain series of frequencies w_j as

$$S_j = (1 + 2 \times \sum_{i=1}^{\infty} r_i \times \cos (w_j \times i)) / \pi$$

where r_i is the autocorrelation of the series at lag i.

Fourier transforms can require a lot of calculation unless special methods are used. The so-called *fast* method takes advantage of the closely related form of the transform at different frequencies. The FOURIER directive in Genstat uses this method.

```
  29  FOURIER [PRINT=transform] Auto; TRANSFORM=Spectrum

29.....................................................................


***** Fourier transform *****
      of sequence Auto

         Unit      Spectrum
            1         2.740
            2         3.558
            3         5.417
            4         3.800
            5         0.771
            6         3.617
            7         2.109
            8         0.784
            9         0.308
           10         0.718
           11         0.381
           12         0.553
```

```
13          0.206
14          0.069
15          0.029
16          0.041
17          0.070
18          0.069
19          0.036
20          0.061
21          0.081
22          0.123
23          0.075
24          0.304
25          0.358
26          0.185
```

The values of this spectrum are calculated at the natural frequencies 0, 1/50, 2/50 ... 25/50, determined by the maximum lag of the autocorrelations: 25. These are not angular frequencies, which some people prefer, but numbers of cycles per unit time; hence 1/50 corresponds to one cycle per 50 years. A given frequency, say f, is not distinguishable from its complementary frequency, $1 - f$: hence the choice of 50 as the divisor, to give the same number of frequencies as autocorrelations. It is not possible to look for cycles with periods shorter than one year (Frequency 1) because there is no information about such behaviour in annual observations.

The choice of maximum lag here can be important to the effectiveness of the analysis. We have used a rule of thumb given in Chatfield (1980, Page 141) and taken twice the square root of the length of the series, which is nearly 25.

The spectrum shows evidence of cyclic behaviour at frequencies near 2/50 and 5/50 (the third and sixth units of the variate Spectrum), corresponding to periods of 25 and 10 years respectively.

11.3 Improving the spectral estimates

The grid of frequencies used by Genstat above by default is rather coarse—the difference between 2/50 and 3/50 corresponds to the difference between 25 and 17 years—so the frequency with largest spectral value is determined quite imprecisely. A grid with more frequencies can be constructed. However, little extra information is gained by narrowing the grid too far, because then the new points simply interpolate smoothly between those that would be calculated for a wider grid. A sensible maximum number of frequencies is four times the maximum lag of the autocorrelations. In Genstat, you can specify the number of frequencies by predefining the length of the variate that is to store the spectrum. The following example shows the spectrum on a grid that is twice as fine as the natural grid.

```
30  VARIATE [VALUES=0, 0.01 ... 0.50] Freqency; DECIMALS=2
31  VARIATE [NVALUES=Freqency] Spectrm2
32  FOURIER [PRINT=transform] Auto; TRANSFORM=Spectrm2

32..................................................................
```

```
***** Fourier transform *****
      of sequence Auto

     Freqency       Spectrm2
         0.00          2.637
         0.01          2.952
         0.02          3.661
         0.03          4.493
         0.04          5.313
         0.05          5.411
         0.06          3.903
         0.07          1.557
         0.08          0.668
         0.09          2.070
         0.10          3.720
         0.11          3.504
         0.12          2.006
         0.13          1.050
         0.14          0.887
         0.15          0.603
         0.16          0.205
         0.17          0.371
         0.18          0.821
         0.19          0.726
         0.20          0.278
         0.21          0.294
         0.22          0.656
         0.23          0.577
         0.24          0.103
         0.25         -0.044
         0.26          0.172
         0.27          0.170
         0.28         -0.074
         0.29         -0.075
         0.30          0.144
         0.31          0.142
         0.32         -0.034
         0.33          0.005
         0.34          0.172
         0.35          0.108
         0.36         -0.067
         0.37         -0.004
         0.38          0.164
         0.39          0.107
         0.40         -0.022
         0.41          0.081
         0.42          0.227
         0.43          0.115
         0.44         -0.028
         0.45          0.138
         0.46          0.407
         0.47          0.407
         0.48          0.255
         0.49          0.241
         0.50          0.288

  33  GRAPH [NROWS=21; NCOLUMNS=61] Spectrm2; Freqency
```

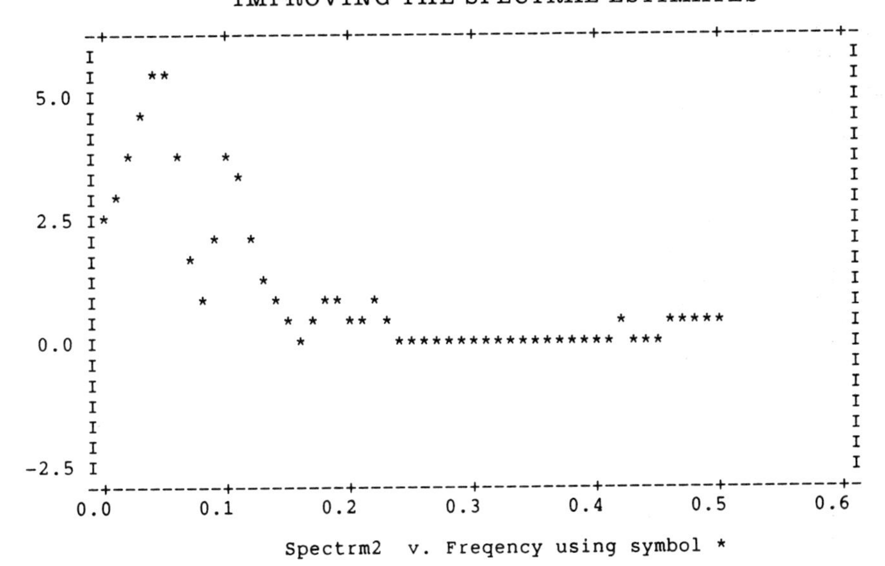

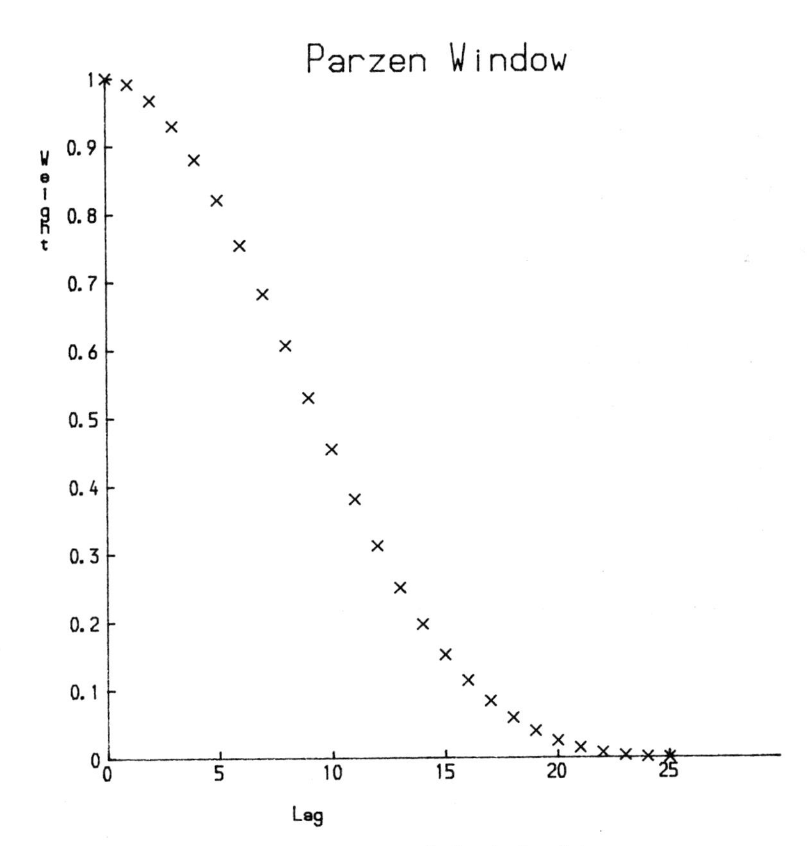

Figure 11.1. 'Parzen window' of weights.

Notice that the printed values of the spectrum are labelled by the corresponding values of the variate Freqency: the FOURIER directive does this if the length of the TRANSFORM variate is declared by reference to another structure, by setting the NVALUES option of VARIATE. Also, the number of decimal places printed for Freqency is defined by the DECIMALS parameter in the VARIATE statement.

The two peaks noticed before can now be estimated as corresponding approximately to frequencies 0.045 and 0.105, equivalent to periods of 22 and 9.5 years.

The estimates of spectral density formed in this way have considerable variance, because they are based on autocorrelations which themselves have large variances. The variances of the spectral values can be reduced by smoothing the autocorrelations before applying the transform—the increase in precision is at the expense of detail in the spectrum. Smoothing is also a useful operation to employ here because a good choice of weights will ensure positive spectrum estimates: unsmoothed autocorrelations can lead to negative estimates. We choose here the "Parzen window" of weights to use in the smoothing, which gives higher weight to autocorrelations at low lags than to those at high lags. Figure 11.1 shows the form of this particular weighting policy.

The weighting can be done with a series of CALCULATE statements. This is shown in the following output, together with the calculation of some extra statistics.

```
34  VARIATE [VALUES=0 ... 25] Weight
35  CALCULATE Weight = Weight / 26
36  & Weight = 2 * (1-Weight)**3 * (Weight.GE.0.5) + \
37      ( 1 - 6 * Weight*Weight*(1-Weight) ) * (Weight.LT.0.5)
38  & Auto = Auto * Weight
39  VARIATE [NVALUES=Freqency] Spectrm3
40  FOURIER Auto; TRANSFORM=Spectrm3
41  GRAPH [NROWS=21; NCOLUMNS=61] Spectrm3; Freqency
```

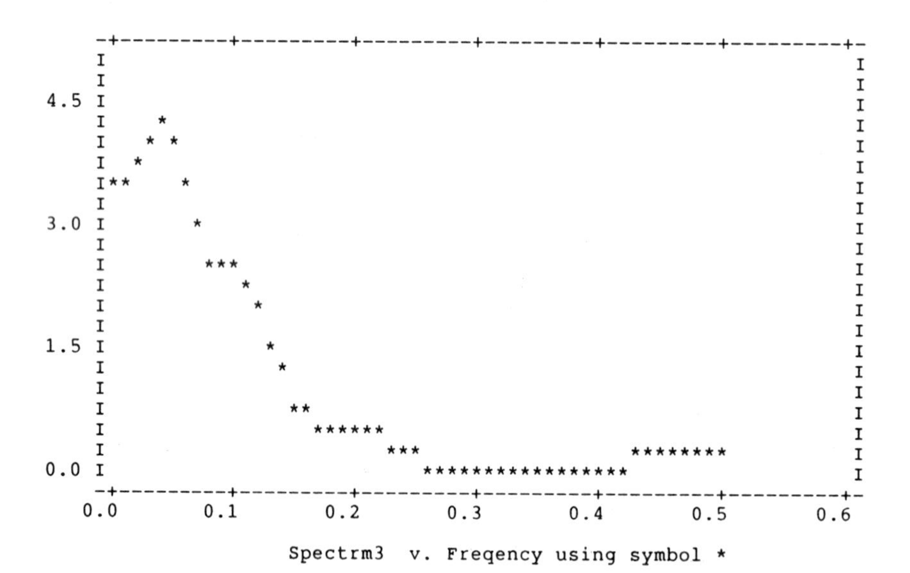

```
42   CALCULATE Temp = NOBSERVATIONS(Diffldday)
43   & Df,Bandwidth = Temp,0.5 * 3.7086 / 26
44   & Upper,Lower = Df / CED(0.025,0.975; Df)
45   PRINT Df,Bandwidth,Lower,Upper; DECIMALS=0,3(3)

         Df      Bandwidt          Lower         Upper
         21         0.071          0.594         2.032
```

The calculations of the additional statistics involve the constant 3.7086, which is the specific value corresponding to the Parzen window. The number of degrees of freedom of the estimates is a guide to the precision of the estimates, which are asymptotically distributed as scaled chi-squared variables. The bandwidth indicates the frequency separation beyond which estimates are effectively independent: the bandwidth decreases if more autocorrelations are used, but so does the number of degrees of freedom. Finally, the 95% confidence limits for the spectrum values are given as multiplying factors. Thus, for example, the fifth spectrum value, at frequency 0.04 or period 25 years, is approximately 4.1 with 95% limits 2.4 (= 4.1 × 0.594) and 8.3 (= 4.1 × 2.032).

It is clear from this spectrum that there is a lot of variance in the series of differenced daylengths that can be associated with a cycle of period about 25 years. Luo *et al.* (1977) point out that 22.3 is the mean period of sunspot polarity fluctuations; therefore there is some support in this analysis for the theory that magnetic solar activity affects the rotation of the Earth through interaction with the Earth's own magnetic field. The other effect noticed in the earlier spectrum is not now so clear; but its period of 9.5 years is also a period found in sunspot activity.

There are many other methods of smoothing that could be used in place of the Parzen window, as there are other ways of deciding how many autocorrelations to use, and at how many frequencies to form the spectral estimates. There is no SPECTRUM directive in Genstat that will make these choices for you; however, it is open to you to write Genstat procedures, as described in Chapter 9, to do spectral analysis in whatever way you choose. The following sections describe the construction of such a procedure, following the choices made in this section.

11.4 A Genstat procedure for spectral analysis

The statements used in the last section to form the smoothed spectrum of the differenced series are given again below.

```
GRAPH Difflday; Year
CORRELATE [MAXLAG = 25; GRAPH = autocorrelations] Difflday; \
    AUTOCORRELATIONS = Auto
VARIATE [VALUES = 0, 0.01 ... 0.50] Freqency; DECIMALS = 2
& [NVALUES = Freqency] Spectrum
& [VALUES = 0...25] Weight
```

```
CALCULATE Weight = Weight / 26
& Weight = 2 * (1 - Weight)**3 * (Weight.GE.0.5) + \
    ( 1 - 6 * Weight*Weight*(1 - Weight) ) * (Weight.LT.0.5)
& Auto = Auto * Weight
FOURIER Auto; TRANSFORM = Spectrum
GRAPH Spectrum; Freqency
CALCULATE Temp = NOBSERVATIONS(Difflday)
& Df,Bandwidt = Temp,0.5 * 3.7086 / 26
& Upper, Lower = Df / CED(0.025, 0.975; Df)
PRINT Df,Bandwidt,Lower,Upper; DECIMALS = 0,3(3)
```

To make these statements work for any series, some of the numbers must be derived from the series. We chose the option setting MAXLAG = 25 according to the rule of thumb that twice the square root of the length of the series is a sensible choice. Therefore, this option setting can be calculated and put in a scalar as follows:

```
CALCULATE Max = 2 * SQRT(NVALUES(Difflday))
```

The value of the scalar will not necessarily be integral. However, the CORRELATE directive will round its value to the nearest integer with no further ado: all Genstat directives that need integers will do this rounding if necessary. The scalar Max can also be used in setting the length and values of the variate Weight.

```
CORRELATE [MAXLAG = Max; GRAPH = autocorrelations] Difflday; \
    AUTOCORRELATIONS = Auto
VARIATE [VALUES = 0 ... Max] Weight
CALCULATE Weight = Weight/(Max + 1)
. . .
```

The frequency values can be set by a similar simple calculation:

```
VARIATE [VALUES = 0, 0.5 ... Max] Freqency
CALCULATE Freqency = Freqency / (Max*2)
```

The discussion of procedures in Chapter 9 gives the information needed to define a procedure to contain the statements above. There are two input structures: the series to be analysed (Difflday) and the variate it is plotted against in the first graph (Year). There is one output structure: the calculated spectrum. These could be specified as parameters of the procedure as follows:

```
PROCEDURE 'Spectral'
    PARAMETER NAME = 'SERIES','TIME','SPECTRUM'; MODE = p
    (statements)
ENDPROCEDURE
```

However, it may often be convenient not to have to specify the time variate, and use the values 1, 2, ... by default. Hence we shall set this variate as an option.

```
PROCEDURE 'Spectral'
    OPTION NAME = 'TIME'; MODE = p; DEFAULT = *
    PARAMETER NAME = 'SERIES','SPECTRUM'; MODE = p
    CALCULATE Nunits = NVALUES(SERIES)
    IF UNSET(TIME)
        VARIATE [VALUES = 1...Nunits] Time
        ASSIGN Time; POINTER = TIME
    ENDIF
    (statements)
ENDPROCEDURE
```

A default structure with values cannot be supplied here unless the procedure is restricted to deal with series of a given length only. This is because a default supplied in an OPTION statement must be the identifier of a given structure. Therefore, we check in the body of the procedure whether the option has been set, as described in Chapter 9. If the option is unset, values are assigned according to the length of the parameter SERIES.

It may also be convenient to have control over which graphs are produced by the procedure. Thus, we introduce a further option which operates like all the PRINT options in standard Genstat directives. The default will be to draw all the graphs. Further, we make the PRINT option the first option of SPECTRAL in the same way as PRINT is first in any Genstat directive that has such an option. The full procedure, including this second option, is given below in a program that stores it in a backing-store file ready for use in later programs, as described in Chapter 9.

```
 1  "Store the SPECTRAL procedure in a backing store file."
 2  PROCEDURE 'Spectral'
 3     OPTION NAME='PRINT','TIME'; MODE=t,p; \
 4        DEFAULT=!T(series,acf,spectrum),*
 5     PARAMETER NAME='SERIES','SPECTRUM'; MODE=p
 6           "Set default times"
 7     CALCULATE Nunits = NVALUES(SERIES)
 8     IF UNSET(TIME)
 9        VARIATE [VALUES=1 ... Nunits] Time
10        DUMMY TIME; VALUE=Time
11     ENDIF
12           "Set default identifier for spectrum"
13     IF UNSET(SPECTRUM)
14        DUMMY SPECTRUM; VALUE=Spectrum
15     ENDIF
16           "Draw graph of series"
17     IF 'series' .IN. PRINT
18        GRAPH [NROWS=21; NCOLUMNS=61] SERIES; TIME
19     ENDIF
20           "Calculate autocorrelations"
21     CALCULATE Max = 2 * SQRT(Nunits)
22     IF 'acf' .IN. PRINT
23        TEXT [VALUES=autocorrelations] Gset
24     ELSE
```

```
25          TEXT [VALUES=*] Gset
26       ENDIF
27       CORRELATE [GRAPH=#Gset; MAXLAG=Max] SERIES; AUTOCORRELATIONS=Auto
28             "Smooth the autocorrelations with the 'Parzen window'"
29       VARIATE [VALUES=0, 0.5 ... Max] Freqency
30       VARIATE [NVALUES=Freqency] SPECTRUM
31       VARIATE [VALUES=0 ... Max] Weight
32       CALCULATE Freqency = Freqency / (Max*2)
33       & Weight = Weight / (Max+1)
34       & Weight = 2 * (1-Weight)**3 * (Weight.GE.0.5) + \
35          ( 1 - 6 * Weight*Weight*(1-Weight) ) * (Weight.LT.0.5)
36       & Auto = Auto * Weight
37             "Form the spectrum"
38       FOURIER Auto; TRANSFORM=SPECTRUM
39       IF 'spectrum' .IN. PRINT
40          GRAPH [NROWS=21; NCOLUMNS=61] SPECTRUM; Freqency
41             "Print the associated statistics"
42          CALCULATE Temp = NOBSERVATIONS(SERIES)
43          & Df,Bandwidth = Temp,0.5 * 3.7086 / (Max+1)
44          & Upper,Lower = Df / CED(.025,0.975;.Df)
45          PRINT Df,Bandwidth,Lower,Upper; DECIMALS=0,3(3)
46       ENDIF
47    ENDPROCEDURE
48    OPEN 'TIMESER.LIB'; CHANNEL=1; FILETYPE=backingstore
49    STORE [CHANNEL=1; PROCEDURE=yes] SPECTRAL
50    STOP
```

11.5 Using the procedure

We shall use the procedure to investigate year-to-year changes in a series of temperatures. The Freshwater Biological Association has recorded the surface temperature of Lake Windermere, in the English Lake District, every weekday since 1933. From *Surface water temperature of Windermere* (Kipling and Roscoe 1977), and later addenda kindly supplied to us by the Freshwater Biological Association, we have taken the number of degree-days in June above 0°C; this is just the sum of the recorded temperatures in °C every day, using the average of Saturday's and Monday's temperature for Sunday. The number of degree-days divided by 30 gives the average June temperature; this series may give evidence of cyclic behaviour in the climate, which would have effects on many plants and animals in the lake.

The program below is all that is required to estimate the spectrum, using as a procedure library the backing-store file set up in the last section.

```
1   "Temperatures on Lake Windermere in June, 1933-84."
2   OPEN 'TIMESER.LIB'; CHANNEL=1; FILETYPE=procedurelibrary
3   VARIATE [VALUES=1933 ... 1984] Year
4   UNITS Year
5   READ [PRINT=data] Degdays

6                   502 459 447 443 454 411 492
7   567 453 418 446 394 424 402 474 432 486
8   503 441 468 483 449 443 441 524 450 508
```

```
 9  535 445 438 480 442 454 504 501 527 503
10  545 438 381 504 486 507 489 466 480 484
11  454 434 503 444 492 :
12  CALCULATE Junetemp = Degdays / 30
13  SPECTRAL [TIME=Year] SERIES=Junetemp; SPECTRUM=Spectrum
```

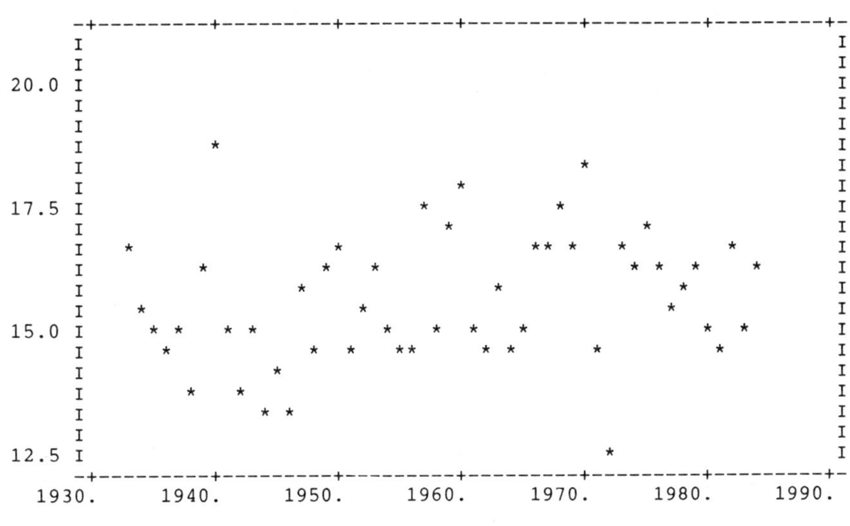

Junetemp v. Year using symbol *

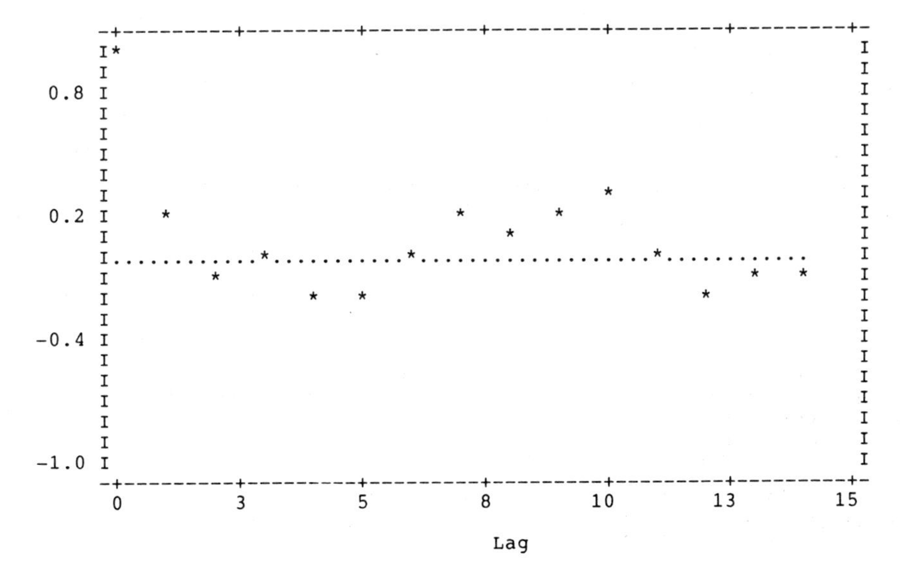

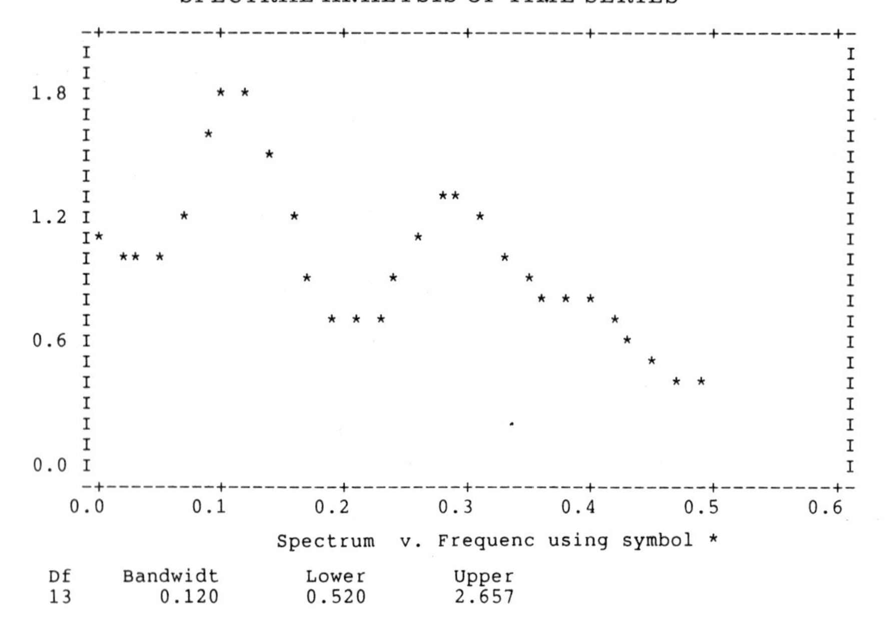

The spectrum shows peaks at about 0.11 and 0.29, corresponding to cycles of period about 9.0 and 3.5 years respectively. However, the confidence limits for the estimates are wide compared to the size of the normalized spectral values at the peaks: the value at 0.11 is about 1.9 with confidence limits 1.0 and 5.0, so we are just 95% confident that this value is greater than the average (1.0). If more data were available, perhaps more information could be extracted. However, it is rare to have records like these for as long as even 52 years: hence the difficulty in many applications of statistics in assessing the effects of climate from year to year.

11.6 Exercises

11(1) Generate a random series of 100 numbers as in Exercise 10(1), and transform them with the NED function so that they form a random Normal series: the NED function transforms uniform numbers in the range (0,1) into Normal deviates.

Compare the spectrum of this *white noise* series with the spectrum of a series derived by adding a trend variate !(0.1,0.2 ... 10) to the white noise series. You do not need to use any smoothing here to see the effect of a trend on the spectrum of a series.

11(2) The following series was analysed in *Genstat 5: an introduction* by the method of harmonic analysis:

98.7	99.1	93.2	97.8	100.2	97.5	101.3
101.3	99.6	107.8	101.4	101.4	100.2	100.9
101.1	102.5	101.0	98.7	99.9	96.7	

The numbers are averages for each month over 12 available 20-month periods of counts of slaughtered sheep in the USA, expressed as percentages of the overall average. Calculate the spectrum of this series, based on autocorrelations up to lag 10, thus giving estimates at the frequencies 0, 1/20, 2/20 ... 10/20. Verify that there is evidence of cyclic behaviour at frequency 1/20—corresponding to a 20-year cycle—as fitted by the first harmonic analysis, and some indication of a 10-year cycle.

11(3) Investigate the cyclic behaviour of the numbers of phone calls recorded between May 1975 and November 1979, shown in Chapter 10, Section 10.2. Compare spectral estimates at frequency 1/12—corresponding to a yearly cycle—calculated (a) without any smoothing, (b) smoothed with the Parzen window, and (c) smoothed with the Tukey window, which has the following weights:

$$\text{Weight}_i = (1 + \cos(\pi i/M))/2 \qquad i = 0,1 \ldots M$$

Appendix Solutions to the exercises

Chapter 1

Exercise 1(1)

```
UNITS [NVALUES = 30]
FACTOR [LEVELS = 3] Block
& [LEVELS = 2; LABELS = !T(Baer,AdeM)] Variety
READ Block,Variety,Rowwidth,Sowdens,Maturity,Yield; \
    FREPRESENTATION = levels,labels,*,*,*,*
    (data)
SCALAR Xl[1,2],Xu[1,2]; VALUE = 145,236,151,242
FOR Varlev = 1,2; Xldummy = Xl[1,2]; Xudummy = Xu[1,2]
    RESTRICT Maturity,Yield; Variety .EQ. Varlev
    MODEL Yield
    FIT Maturity
    RKEEP FITTEDVALUES = Fyield
    GRAPH [YTITLE = 'Grain yield - g - '; XTITLE = 'Days to maturity'; \
        YLOWER = 400; YUPPER = 720; XLOWER = Xldummy; \
        XUPPER = Xudummy; NROWS = 25; NCOLUMNS = 61] \
        Fyield,Yield; Maturity; METHOD = line,point
ENDFOR
RESTRICT Maturity,Yield,Fyield,Variety
MODEL Yield
FIT [PRINT = summary,estimates,accumulated] Variety * Maturity
ENDJOB
```

Exercise 1(2)

```
UNITS [NVALUES = 30]
FACTOR [LEVELS = 3] Block
& [LEVELS = 2; LABELS = !T(Baer,AdeM)] Variety
READ Block,Variety,Rowwidth,Sowdens,Maturity,Yield; \
    FREPRESENTATION = levels,labels,*,*,*,*
    (data)
CALCULATE RS = Rowwidth * Sowdens
```

```
& S[2] = Sowdens ** 2
TEXT [VALUES = 'Days to maturity'] Yt[1]
& [VALUES = 'Yield - g - '] Yt[2]
FOR Y = Maturity,Yield; Ytdummy = Yt[1,2]
    MODEL Y
    TERMS Block + Variety * (Rowwidth + Sowdens + RS + S[2])
    FIT [PRINT = summary,estimates,accumulated] \
        Block + Variety * (Rowwidth + Sowdens + RS + S[2])
    DROP [PRINT = *] Block
    RKEEP ESTIMATES = Esti
    DROP Rowwidth + Variety.Rowwidth
    SCALAR B[1...10]
    CALCULATE B[1...10] = Esti$[1...10]
    UNITS [NVALUES = 21]
    VARIATE [VALUES = 10...30] Sdrang
    CALCULATE Fy[1,2] = B[1] + B[2] * 0.4,0.2 + \
        B[3] * Sdrang + B[4] * 0.4,0.2 * Sdrang + \
        B[5] * Sdrang ** 2
    CALCULATE Fy[3,4] = Fy[1,2] + B[6] + B[7] * 0.4,0.2 + \
        B[8] * Sdrang + B[9] * 0.4,0.2 * Sdrang + \
        B[10] * Sdrang ** 2
    GRAPH [YTITLE = Ytdummy; XTITLE = 'Sowing density (kg/ha)'] \
        Fy[1...4],Y; 4(Sdrang),Sowdens; METHOD = 4(line),point
ENDFOR
STOP
```

It is not possible to conduct equivalent analyses using the BLOCKS, TREATMENTS, and ANOVA directives because the highest sowing density does not occur in combination with the greater row width, which makes the design unbalanced.

Exercise 1(3)

```
UNITS [NVALUES = 8]
READ Solvent,AmtC,ConcC,Time,AmtB,Yield
    (data)
CALCULATE SA = Solvent * AmtC
& ST = Solvent * Time
MODEL Yield
TERMS Solvent + AmtC + ConcC + Time + AmtB + SA + ST
FIT Solvent + AmtC + ConcC + Time + AmtB
RKEEP ESTIMATES = Esti
SCALAR B[1...6]
CALCULATE B[1...6] = Esti$[1...6]
```

```
MATRIX [ROWS = 4; COLUMNS = 6; VALUES = (200,210...250)4] Solvrang
& [VALUES = 6(90...93)] CCrang
CALCULATE Fyield = B[1] + B[2]*Solvrang + B[4]*CCrang
CONTOUR Fyield
DROP Solvent
ADD Solvent + SA
& ST
STOP
```

Chapter 2

Exercise 2(1)

```
UNITS [NVALUES = 9]
FACTOR [LABELS = !T(I,II,III)] Soil
& [LABELS = !T(Owned,Rented,Mixed)] Tenure
GENERATE Soil,Tenure
TABLE [CLASSIFICATION = Soil,Tenure] Tfarmers
READ Tfarmers
    36   67   49   31   60   49   58   87   80   :
VARIATE Farmers; VALUES = Tfarmers
MARGIN Tfarmers
PRINT Tfarmers; DECIMALS = 0
MODEL [LINK = log; DISTRIBUTION = poisson] Farmers
FIT Soil + Tenure
STOP
```

Exercise 2(2)

```
UNITS [NVALUES = 8]
FACTOR [LABELS = !T(Cut,Entire)] Leafshap
& [LABELS = !T(Green,Yellow)] Leafcolr
& [LABELS = !T(Purple,Green)] Stemcolr
GENERATE Leafshap,Leafcolr,Stemcolr
READ [SERIAL = yes] Seedlngs,Logprob
    57   5   15   9   13   11   1   7:
    27   9   9    3   9    3    3   1:
CALCULATE Logprob = LOG(Logprob / SUM(Logprob))
MODEL [LINK = log; DISTRIBUTION = poisson; OFFSET = Logprob] \
    Seedlngs
TERMS Leafshap*Leafcolr*Stemcolr
```

```
FIT [PRINT = *]
ADD [PRINT = *] Leafshap
& Leafcolr
& Stemcolr
& Leafshap.Leafcolr
& Leafshap.Stemcolr
& [PRINT = model,accumulated] Leafcolr.Stemcolr
TABULATE [CLASSIFICATION = Leafshap,Leafcolr; PRINT = totals] \
    Seedlngs
& [CLASSIFICATION = Leafshap,Stemcolr] Seedlngs
& [CLASSIFICATION = Leafcolr,Stemcolr] Seedlngs
STOP
```

Exercise 2(3)

```
UNITS [NVALUES = 18]
FACTOR [LEVELS = 9] Ageraw
& [LEVELS = 2; LABELS = !T(Male,Female)] Sex
GENERATE Ageraw,Sex
READ [SERIAL = yes] Gtraw,Lpraw[1,2]
    83 67 31 31 16 7 3 2 0 92 37 22 12 6 3 0 0 0 :
    0.2606 0.1440 0.0796 0.0440 0.0243 0.0134 0.0074 0.0041 0.0051
    0.1868 0.1032 0.0570 0.0315 0.0174 0.0096 0.0053 0.0029 0.0036 :
    0.2357 0.1403 0.0835 0.0497 0.0296 0.0176 0.0105 0.0062 0.0092
    0.2189 0.1041 0.0495 0.0236 0.0112 0.0053 0.0025 0.0012 0.0011 :
FACTOR [LEVELS = 6] Age
CALCULATE Age = NEWLEVELS(Ageraw; !(1...6,6,6,6))
VARIATE [NVALUES = 12] Gttits,Logprob[1,2]
FOR In = Gtraw,Lpraw[1,2]; Out = Gttits,Logprob[1,2]
    TABULATE [CLASSIFICATION = Age,Sex] In; TOTALS = T
    EQUATE OLD = T; NEW = Out
ENDFOR
CALCULATE Logprob[1,2] = LOG(Logprob[1,2])
FOR Dlogprob = Logprob[1,2]
    MODEL [LINK = log; DISTRIBUTION = poisson; OFFSET = Dlogprob] \
        Gttits
    FIT [PRINT = model,summary]
ENDFOR
STOP
```

Genstat cannot give the correct number of degrees of freedom because it does not know that the values of the offset variate depend on either one or two values obtained from the data, namely the expected proportion or proportions dying. The residual number of degrees of freedom should therefore be reduced from 11 to 10 in the first analysis, and to 9 in the second. The deviance explained by allowing the expected proportion to vary is the difference between the residual deviances; that is, $27.53 - 18.14 = 9.39$ with one degree of freedom.

Chapter 3

Exercise 3(1)

```
VARIATE [NVALUES = 16] Latex
& [VALUES = 1,3...31] Week
READ Latex
    .776   .852   .850   .869   .939   .904   .930   .948   .942   .938
    .979   .975   .955   .993   .985   1.013 :
MODEL Latex
FITCURVE Week
STOP
```

Exercise 3(2)

```
UNITS [NVALUES = 15]
READ Water
     1.3    1.3    1.9    3.4    5.3    7.1    10.6   16.0
    16.4   18.3   20.9   20.5   21.3   21.2   20.9 :
VARIATE [VALUES = 0.5...14.5] Distance
MODEL Water
FITCURVE [CURVE = logistic] Distance
RKEEP FITTEDVALUES = Fit; RESIDUALS = Res
GRAPH Fit,Water; Distance; METHOD = line,point
& Res; Fit
CALCULATE W = 1/Water
MODEL [WEIGHTS = W] Water
FITCURVE [CURVE = logistic] Distance
RKEEP FITTEDVALUES = Fit; RESIDUALS = Res
GRAPH Res; Fit
FITCURVE [CURVE = glogistic] Distance
STOP
```

Exercise 3(3)

```
UNITS [NVALUES = 10]
READ Yield
```

```
    232.65    104.75
    369.08    188.63
    455.63    211.75
    491.45    217.63
    511.50    231.13 :
VARIATE [VALUES = 2(0...4)] Fertiliz
FACTOR [LEVELS = !(1945,1946); VALUES = (1945,1946)5] Year
MODEL Yield
TERMS Fertiliz*Year
FITCURVE [PRINT = *] Fertiliz
ADD [PRINT = *] Year
& Fertiliz.Year
& [PRINT = model,summary,estimates,accumulated; \
    NONLINEAR = separate]
STOP
```

Chapter 4

Exercise 4(1)

```
UNITS [NVALUES = 8]
FACTOR [LEVELS = !(200,250)] Solvent
& [LEVELS = !(4.0,4.5)] AmtC
& [LEVELS = !(90,93)] ConcC
& [LEVELS = !(1,2)] Time
& [LEVELS = !(3.0,3.5)] AmtB
READ Solvent,AmtC,ConcC,Time,AmtB,Yield
    (data)
TREATMENTS Solvent + AmtC + ConcC + Time + AmtB
ANOVA [FPROBABILITY = yes] Yield
TREATMENTS Solvent * AmtC * ConcC * Time * AmtB
ANOVA [FPROBABILITY = yes; FACTORIAL = 5] Yield
STOP
```

The interactions that are described as aliased in the information summary produced by the ANOVA statement correspond to the product terms that cannot be fitted in a regression model. For example, the Solvent.Time interaction would be estimated by the contrast of observations 1, 4, 5, and 8 with observations 2, 3, 6, and 7, but this contrast has already been used to estimate the main effect of AmtB. In order to estimate all the main effects and interactions of five factors at two levels, at least $2^5 = 32$ observations would be needed.

Exercise 4(2)

```
UNITS [NVALUES = 48]
FACTOR [LEVELS = 3; VALUES = 16(1...3)] Location
& [LEVELS = 4; \
    LABELS = !T('Alder 1','Alder 2','Willow 1','Willow 2'); \
    VALUES = (1...4)12] Species
READ Growth
    (data)
MATRIX [ROWS = 3; COLUMNS = 4; VALUES = -1,-1,+1,+1, \
                                        -1,+1, 0, 0, \
                                         0, 0,-1,+1] Coef
TREATMENTS Location * REG(Species; 3; Coef)
ANOVA [FPROBABILITY = yes] Growth
BLOCKS Location / Species
TREATMENTS REG(Species; 3; Coef)
ANOVA [FPROBABILITY = yes] Growth
STOP
```

Exercise 4(3)

```
UNITS [NVALUES = 108]
FACTOR [LEVELS = !(-4,-1,0,1); VALUES = 27(-4,-1,0,1)] \
    Injctday
& [LEVELS = 3; VALUES = 9(1...3)4] Batch
& [LEVELS = 2; LABELS = !T(Treated,Control); \
    VALUES = (6(1),3(2))12] Tc
READ Plaques
    (data)
CALCULATE Tplaques = Plaques ** 0.5
TREATMENTS (POL(Injctday; 3) / Batch) * Tc
ANOVA [FPROBABILITY = yes] Plaques,Tplaques; \
    FITTEDVALUES = Fp,Ftp; RESIDUALS = Rp,Rtp
FOR R = Rp,Rtp; F = Fp,Ftp
    GRAPH R; F
ENDFOR
FOR Invar = Plaques,Tplaques; Outvar = Diff,Tdiff
    TABULATE [CLASSIFICATION = Tc,Injctday,Batch] Invar; MEANS = T
    VARIATE [NVALUES = 24] V
    EQUATE OLD = T; NEW = V
    CALCULATE Outvar = V$[!(1...12)] - V$[!(13...24)]
ENDFOR
```

```
FACTOR [MODIFY = yes; NVALUES = 12] Injctday
GENERATE Injctday
TREATMENTS POL(Injctday; 3)
ANOVA [FPROBABILITY = yes] Diff,Tdiff
STOP
```

Chapter 5

Exercise 5(1)

```
UNITS [NVALUES = 26]
POINTER [VALUES = K,Cl,S,Si,Mg] Elements
READ Elements[]
    (data)
MATRIX [ROWS = 26; COLUMNS = 2] Scores
PCP [PRINT = roots,loadings,scores] Elements; SCORES = Scores
VARIATE [NVALUES = 26] Score[1,2]
CALCULATE Score[] = Scores$[*; 1,2]
FACTOR [LEVELS = 26; VALUES = 1...26] Labels
GRAPH [TITLE = 'PCP scores for 26 insects'; YTITLE = 'PCP Score 2'; \
    XTITLE = 'PCP Score 1'; EQUAL = scale; NROWS = 37; \
    NCOLUMNS = 61] Score[2]; Score[1]; SYMBOLS = Labels
```

Exercise 5(2)

```
POINTER [NVALUES = 5] LogData
CALCULATE LogData[] = LOG10(Elements[])
LRV [ROWS = Elements; COLUMNS = 2] Lrv
PCP [PRINT = roots,loadings,scores] LogData; LRV = Lrv; \
    SCORES = Scores
CALCULATE Score[] = Scores$[*; 1,2]
VARIATE [NVALUES = 5] Load[1,2]
CALCULATE Load[] = Lrv[1]$[*; 1,2]
TEXT [VALUES = K,Cl,S,Si,Mg] Elabels
GRAPH [TITLE = 'Biplot from logs of 5 elements for 26 insects'; \
    YTITLE = 'Axis 2'; XTITLE = 'Axis 1'; EQUAL = scale; NROWS = 37; \
    NCOLUMNS = 61] Score[2],Load[2]; Score[1],Load[1]; \
    SYMBOLS = Labels,Elabels
PCP [PRINT = roots,loadings,scores; METHOD = correlation] Elements; \
    LRV = Lrv; SCORES = Scores
CALCULATE Score[] = Scores$[*; 1,2]
```

```
CALCULATE Load[] = Lrv[1]$[*; 1,2]
GRAPH [TITLE = 'Biplot from correl.s of 5 elements for 26 insects'; \
    YTITLE = 'Axis 2'; XTITLE = 'Axis 1'; EQUAL = scale; NROWS = 37; \
    NCOLUMNS = 61] Score[2],Load[2]; Score[1],Load[1]; \
    SYMBOLS = Labels,Elabels
```

Exercise 5(3)

```
FACTOR [VALUES = (1)9,(2)8,(3)9; LABELS = !T(a,b,c)] Group
SSPM [TERMS = LogData[]; GROUPS = Group] Wsspm
FSSPM Wsspm
TEXT [VALUES = A,B,C] Glabels
MATRIX [ROWS = Glabels; COLUMNS = 2] GScores
CVA [PRINT = roots,loadings] Wsspm; LRV = Lrv; SCORES = GScores
VARIATE [NVALUES = Group] GScore[1,2]
MATRIX [ROWS = 26; COLUMNS = 5] Data
CALCULATE Data$[*; 1...5] = LogData[]
& Scores = Data *+ Lrv[1]
& Score[] = Scores$[*; 1,2]
& Score[] = Score[] - MEAN(Score[])
& GScore[] = GScores$[*; 1,2]
PRINT Group,Score[]
& Glabels,GScore[]
GRAPH [YTITLE = 'CVA Score 2'; XTITLE = 'CVA Score 1'; \
    TITLE = 'CVA for 3 groups from logs of 5 elements for 26 insects'; \
    EQUAL = scale; NROWS = 37; NCOLUMNS = 61] \
    Score[2],GScore[2]; Score[1],GScore[1]; SYMBOLS = Group,Glabels
STOP
```

Chapter 6

Exercise 6(1)

```
TEXT [VALUES = 'Long Ashton',Rothamsted,Henlow, \
    Bayfordbury,Norwich] Places
SYMMETRICMATRIX [ROWS = Places] Distance
READ Distance
     0
    70     0
    46    66     0
    76   102    34     0
    72    81    19    50     0 :
```

```
CALCULATE Distance = - Distance * Distance / 2
LRV [ROWS = Places; COLUMNS = 3] Lrv; VECTORS = Coords
PCO [PRINT = roots,scores,centroid,residuals; NROOTS = 3] Distance; \
    LRV = Lrv
MATRIX [ROWS = Places; COLUMNS = 2] True
READ True
    354    171
    514    204
    531    208
    623    308
    518    238 :
ROTATE [PRINT = rotations,coordinates,residuals,sums; SCALING = yes; \
    STANDARDIZE = centre] True; Coords
STOP
```

Exercise 6(2)

```
TEXT [NVALUES = 16] Station
VARIATE [NVALUES = 16] Truex,Truey,Mapx,Mapy
READ Station,Truex,Truey,Mapx,Mapy
    'Oxford Circus'            73    61    41    36
    'Bond Street'              44    53    24    36
    'Regent''s Park'           52   111    41    43
    'Tottenham Court Road'    118    68    60    36
    'Covent Garden'           142    47    81    17
    'Leicester Square'        126    37    60    17
    'Piccadilly Circus'       103    27    47    17
    'Green Park'               71     6    23    17
    'Marble Arch'              10    47     6    36
    'Baker Street'             15   104    24    63
    'Great Portland Street'    64   114    32    60
    'Warren Street'            85   120    60    56
    'Goodge Street'           101    94    60    46
    'Holborn'                 158    76    91    34
    'Aldwych'                 174    44    78    14
    'Charing Cross'           135    16    60     3 :
  PRINT Station,Truex,Truey,Mapx,Mapy; DECIMALS = 0
MATRIX [ROWS = Station; COLUMNS = 2] True,Map
& [COLUMNS = 1] Res
CALCULATE True$[*; 1,2] = Truex,Truey
& Map$[*; 1,2] = Mapx,Mapy
    " Scale the actual locations to be measured in yards. "
```

```
& True = 1760 * True / 90
ROTATE [PRINT = rotations,coordinates,residuals,sums; SCALING = yes; \
    STANDARDIZE = centre] True; Map; \
    XOUTPUT = True; YOUTPUT = Map; RESIDUALS = Res
CALCULATE Truex,Truey = True$[*; 1,2]
& Mapx,Mapy = Map$[*; 1,2]
TEXT Real,OnMap; VALUES = \
    !T(OC,BS,RP,TC,CG,LS,PC,GP,MA,BS,GP,WS,GS,H,A,CC), \
    !T(oc,bs,rp,tc,cg,ls,pc,gp,ma,bs,gp,ws,gs,h,a,cc)
GRAPH [EQUAL = scale; NROWS = 37; NCOLUMNS = 61; \
    TITLE = 'True locations, and those from Map, for 16 Stations'] \
    Truey,Mapy; Truex,Mapx; SYMBOLS = Real,OnMap
VARIATE ResYards; VALUES = Res
PRINT Station,Real,OnMap,ResYards; FIELDWIDTH = *,10,6,10; \
    DECIMALS = 0; JUSTIFICATION = left,(right)3
STOP
```

Exercise 6(3)

```
UNITS [NVALUES = 26]
POINTER [VALUES = K,Cl,S,Si,Mg] Elements
READ Elements[]
    (data)
POINTER [NVALUES = 5] LogData
CALCULATE LogData[] = LOG10(Elements[])
LRV [ROWS = 26; COLUMNS = 5] Lrvlog
PCO [PRINT = roots,scores] LogData; LRV = Lrvlog
MATRIX [ROWS = 26; COLUMNS = 5] Scorecor
PCP [PRINT = roots,scores; METHOD = correlation] Elements; \
    SCORES = Scorecor
ROTATE [PRINT = rotations,coordinates,residuals,sums] \
    Lrvlog[1]; Scorecor
STOP
```

Chapter 7

Exercise 7(1)

```
UNITS [NVALUES = 26]
POINTER [VALUES = K,Cl,S,Si,Mg] Elements
READ Elements[]
    (data)
```

```
    " To avoid extreme values causing problems, a log
    transformation can be used. Either a city-block or a
    Pythagorean method of forming similarity is appropriate. "
FSIMILARITY [SIMILARITY = Sim] Elements[]; TEST = 3
HCLUSTER [PRINT = dendrogram,amalgamations; \
    METHOD = averagelink] Sim
```

Exercise 7(2)

```
HCLUSTER [PRINT = dendrogram,amalgamations; \
    METHOD = singlelink] Sim
LRV [ROWS = 26; COLUMNS = 2] PCOLrv
PCO Sim; LRV = PCOLrv
VARIATE [NVALUES = 26] Score[1,2]
CALCULATE Score[] = PCOLrv[1]$[*; 1,2]
FACTOR [LEVELS = 26; VALUES = 1...26] Ulabel
FOR Title = \
    'MST links: > 88% (full lines); > 85% (dashed lines)', \
    'MST links: > 85% (full lines); > 80% (dashed lines)', \
    'MST links: > 80% (full lines); > 75% (dashed lines)'
    GRAPH [TITLE = Title; YTITLE = 'Principal coordinate 2'; \
        XTITLE = 'Principal coordinate 1'; EQUAL = scale; NROWS = 37; \
        NCOLUMNS = 61] Score[2]; Score[1]; SYMBOLS = Ulabel
ENDFOR
```

Exercise 7(3)

```
HCLUSTER [METHOD = averagelink] Sim; GTHRESHOLD = 65; \
    GROUPS = Avlinkge
FACTOR [LEVELS = 3; VALUES = (1)9,(2)8,(3)9] APriori
PRINT [ORIENTATION = across; RLWIDTH = 10] Ulabel,APriori,Avlinkge; \
    FIELDWIDTH = 4; DECIMALS = 0
TABULATE [PRINT = counts; CLASSIFICATION = Avlinkge,APriori; \
    MARGINS = yes]
STOP
```

Chapter 8

Exercise 8(1)

```
JOB 'Fitting circles'
    " This Job takes the (x,y) coordinates of a set of points and
    finds either the least-squares estimates, or the least-squares
```

```
    squared estimates, of the parameters of the best-fitting
    circle to those points. The parameters are the centre of the
    circle (a_hat,b_hat) and the radius (r_hat). "
SCALAR Method,Lots,Thresh,Report,Graphopt; VALUE = 2,100,0.0001,1,1
OPEN 'Circle.dat','Circle.lis'; CHANNEL = 2; FILETYPE = input,output
READ [CHANNEL = 2; SETNVALUES = yes] X,Y
CLOSE 2; FILETYPE = input
POINTER [VALUES = A_hat,B_hat] Centre
SCALAR Centre[],R_hat,Rss,Oldrss,Change,Iterno; VALUE = 0
VARIATE [NVALUES = X] Xadj,Yadj,R
CALCULATE Centre[] = MEAN(X,Y)
IF Method == 1
    CALCULATE Oldrss = SUM(X**2 + Y**2)
    IF Report
        PRINT [CHANNEL = 2; IPRINT = *; ORIENTATION = across] \
            !T(iteration,A_hat,B_hat,R_hat,Residual,Criterion); \
            FIELDWIDTH = 12
        & [SQUASH = yes] Iterno,Centre[]; \
            FIELDWIDTH = 12; DECIMALS = 0,(*)2
    ENDIF
    FOR [NTIMES = Lots]
        CALCULATE Iterno = Iterno + 1
        & Xadj,Yadj = X,Y - Centre[]
        & R_hat = MEAN(R = SQRT(Xadj**2 + Yadj**2))
        & Rss = SUM((R - R_hat) ** 2)
        & [ZDZ = zero] R = (R > 0) / R
        & Centre[] = SUM(X,Y * (1 - R)) / SUM(1 - R)
        & Change = ABS(Oldrss - Rss) / (R_hat ** 2)
        IF Report
            PRINT [CHANNEL = 2; IPRINT = *; SQUASH = yes] \
                Iterno,Centre[],R_hat,Rss,Change; \
                FIELDWIDTH = 12; DECIMALS = 0,(*)5
        ENDIF
        EXIT Change < Thresh
        CALCULATE Oldrss = Rss
    ENDFOR
ELSIF Method == 2
    CALCULATE Xadj,Yadj = X,Y - Centre[]
    & R = Xadj * Xadj + Yadj * Yadj
    & Centre[] = SUM(Yadj,Xadj*Yadj,Xadj)*SUM(Xadj,Yadj*R) \
        - SUM(Xadj*Yadj)*SUM(Yadj,Xadj*R)
```

```
    & Centre[] = MEAN(X,Y) + Centre[]/(2*(SUM(Xadj*Xadj) \
        *SUM(Yadj*Yadj) - (SUM(Xadj*Yadj))**2))
    & R_hat = SQRT((MEAN(X) - A_hat)**2 \
        + (MEAN(Y) - B_hat)**2 + MEAN(R))
ELSE
    PRINT [IPRINT = *] 'Invalid value of Method',Method
ENDIF
FOR
    IF Method == 2
        PRINT [CHANNEL = 2] 'Least-squares squared solution - '
        & Centre[],R_hat; FIELDWIDTH = 12
    ELSIF Change < Thresh
        PRINT [CHANNEL = 2; IPRINT = *] \
            'Algorithm converged after',Iterno,' iterations'; \
            FIELDWIDTH = 4; DECIMALS = 0
        PRINT [CHANNEL = 2] 'Final solution - '
        & Centre[],R_hat
        & [IPRINT = *] 'with residual sum of squares',Rss
    ELSIF Report
        PRINT [CHANNEL = 2] '*** Algorithm failed to converge ***'
    ELSE
        PRINT '*** Algorithm failed to converge ***'
        & Iterno,Centre[],R_hat,Rss,Change; \
            FIELDWIDTH = 12; DECIMALS = 0,(*)5

    ENDIF
    IF ((Method == 2) .OR. (Change < Thresh)) * Graphopt
        VARIATE [VALUES = 0,15...360] Angle
        CALCULATE Circlex = COS(Angle*6.28318/360)*R_hat + A_hat
        & Circley = SIN(Angle*6.28318/360)*R_hat + B_hat
        CASE Graphopt
            GRAPH [CHANNEL = 2; TITLE = 'Points and fitted circle'; \
                JOIN = given; EQUAL = scale; NROWS = 37; \
                NCOLUMNS = 61] B_hat,Circley,Y; A_hat,Circlex,X; \
                METHOD = point,curve; SYMBOLS = ' + ','*','S'
        OR
            OPEN 'Circle.grd'; CHANNEL = 1; FILETYPE = graphics
            FRAME 1; YLOWER = 0; YUPPER = 1; XLOWER = 0; XUPPER = 1
            AXES [EQUAL = scale] 1; STYLE = none
            PEN 1...3; COLOUR = 1; METHOD = point,closed; \
                LINESTYLE = 0,1; SYMBOLS = 1,0,4; JOIN = given
            DGRAPH [TITLE = 'Points and fitted circle'; WINDOW = 1; \
```

```
                KEYWINDOW = 0] B_hat,Circley,Y; A_hat,Circlex,X; \
                PEN = 1...3
        ELSE
            PRINT [IPRINT = *] 'Invalid value of Graphopt',Graphopt
        ENDCASE
    ENDIF
ENDFOR
STOP
```

Exercise 8(2)

```
SCALAR Stats,Hists,Nvars; VALUE = 1,1,*
OPEN 'Sol82.dat'; CHANNEL = 2
READ [CHANNEL = 2] Nvars
POINTER [NVALUES = Nvars] V
READ [CHANNEL = 2; SETNVALUES = yes] V[]
CLOSE 2
FOR
    IF Stats
        POINTER [NVALUES = Nvars] Nobsrvtn,Nmissing,Mean,Variance, \
            Median,Minimum,Maximum,Range,Stnd_Dev
        CALCULATE Nunits = NVALUES(V[1])
        & Nobsrvtn[] = NOBSERVATIONS(V[])
        & Nmissing[] = NMV(V[])
        & Mean[] = MEAN(V[])
        & Variance[] = VARIANCE(V[])
        & Median[] = MEDIAN(V[])
        & Minimum[] = MINIMUM(V[])
        & Maximum[] = MAXIMUM(V[])
        & Range[] = Maximum[] - Minimum[]
        & Stnd_Dev[] = SQRT(Variance[])
        PRINT [IPRINT = *] 'Summary information for',Nvars, \
            ' variables of length',Nunits; FIELDWIDTH = *,2; DECIMALS = 0
        & [ORIENTATION = across] !T(Variable,'No.Observed', \
            Missing,Mean,Variance,Median,Minimum,Maximum, \
            Range,StandardDev); FIELDWIDTH = 12
        FOR Dnobsrvtn = Nobsrvtn[]; Dnmissing = Nmissing[]; \
            Dmean = Mean[]; Dvariance = Variance[]; \
            Dmedian = Median[]; Dminimum = Minimum[]; \
            Dmaximum = Maximum[]; Drange = Range[]; \
            Dstnd_dev = Stnd_Dev[]; Varno = 1...Nvars
```

```
            PRINT [IPRINT = *; SQUASH = yes] Varno,Dnobsrvtn, \
                Dnmissing,Dmean,Dvariance,Dmedian,Dminimum, \
                Dmaximum,Drange,Dstnd_dev
        ENDFOR
    ENDIF
    IF Hists
        FOR Dv = V[]
            HISTOGRAM Dv
        ENDFOR
    ENDIF
ENDFOR
```

Exercise 8(3)

```
SCALAR Graphs; VALUE = 1
IF Graphs
    FOR Dy = V[]
        FOR Dx = V[]
            EXIT Dy .IS. Dx
            GRAPH Dy; Dx
        ENDFOR
    ENDFOR
ENDIF
STOP
```

Chapter 9

Exercise 9(1)

```
PROCEDURE 'BARTLETT'
    OPTION NAME = 'PRINT'; MODE = t; DEFAULT = 'stats'
    PARAMETER NAME = 'SS','DF','CHISQD','DFCHISQD'; MODE = p
    IF UNSET(CHISQD)
        ASSIGN Chisqd; POINTER = CHISQD
    ENDIF
    IF UNSET(DFCHISQD)
        ASSIGN DFchisqd; POINTER = DFCHISQD
    ENDIF
    SCALAR DFCHISQD,Logavvar,Smlogvar,CHISQD,Corrfac
    CALCULATE DFCHISQD = NVALUES(DF)
    UNITS [NVALUES = DFCHISQD]
```

```
    CALCULATE DFCHISQD = DFCHISQD - 1
    & Logavvar = LOG(SUM(SS) / SUM(DF))
    & Var = SS / DF
    & Smlogvar = SUM(DF * LOG(Var))
    & CHISQD = Logavvar * SUM(DF) - Smlogvar
    & Corrfac = SUM(1 / DF) - (1 / SUM(DF))
    & Corrfac = 1 + (1 / (3 * DFCHISQD)) * Corrfac
    & CHISQD = CHISQD / Corrfac
    IF 'stats' .IN. PRINT
        PRINT 'Bartlett''s Test for Homogeneity of Variance'
        & CHISQD,DFCHISQD; DECIMALS = 3,0
    ENDIF
ENDPROCEDURE

VARIATE [VALUES = 27.4,18.7,12.1,4.2,9.8] ResSS
& [VALUES = 4,7,10,2,5] ResDF
BARTLETT [PRINT = stats] SS = ResSS; DF = ResDF
BARTLETT SS = ResSS; DF = ResDF
SCALAR Barttest,BartDF
BARTLETT [PRINT = nothing] SS = ResSS; DF = ResDF; \
    CHISQD = Barttest; DFCHISQD = BartDF
PRINT Barttest,BartDF; DECIMALS = 3,0
STOP
```

Exercise 9(2)

```
PROCEDURE 'TIMECONVERT'
    OPTION NAME = 'SECONDS','CLOCK'; MODE = t; \
        DEFAULT = 'present','12-hour'
    PARAMETER NAME = 'HOUR','MINUTE','SECOND','AMPM','RESULT'; \
        MODE = p
    UNITS HOUR
    VARIATE RESULT
    CALCULATE RESULT = 0
    IF '12-hour' .IN. CLOCK
        FACTOR [LEVELS = !(0,12)] Ampmhrs
        CALCULATE Ampmhrs = NEWLEVELS(AMPM; !(0,12))
        CALCULATE RESULT = RESULT + Ampmhrs
    ENDIF
    CALCULATE RESULT = RESULT + HOUR
    & RESULT = RESULT * 60 + MINUTE
    IF 'present' .IN. SECONDS
```

```
        CALCULATE RESULT = RESULT * 60 + SECOND
    ENDIF
ENDPROCEDURE

UNITS [NVALUES = 5]
FACTOR [LABELS = !T(AM,PM)] Ampm
READ Hour,Minute,Ampm; FREPRESENTATION = label
     9 20 AM
     5 31 AM
     2 40 PM
     1 12 AM
    11 15 PM:
TIMECONVERT [SECONDS = absent] HOUR = Hour; MINUTE = Minute; \
    AMPM = Ampm; RESULT = Timemins
PRINT Timemins; DECIMALS = 0
UNITS [NVALUES = 5]
READ Hour,Minute,Second
     9 20  5
     5 31 34
    14 40 21
     1 12 56
    23 15 12:
TIMECONVERT [CLOCK = '24-hour'] HOUR = Hour; MINUTE = Minute; \
    SECOND = Second; RESULT = Timesecs
PRINT Timesecs; DECIMALS = 0
STOP
```

Exercise 9(3)

```
PROCEDURE 'HEATUNIT'
    OPTION NAME = 'NEGSZERO','NUMBER','DURATION','PRINT'; \
        MODE = t; DEFAULT = 'no','no','no','tempdiff'
    PARAMETER NAME = 'TEMP','CRITT','MEANDIFF','NPERIODS', \
        'DPERIODS'; MODE = p
    IF UNSET(MEANDIFF)
        ASSIGN Meandiff; POINTER = MEANDIFF
    ENDIF
    IF UNSET(NPERIODS)
        ASSIGN Nperiods; POINTER = NPERIODS
    ENDIF
    IF UNSET(DPERIODS)
        ASSIGN (Dperiods); POINTER = DPERIODS
```

```
    ENDIF
    UNITS TEMP
    CALCULATE Tempdiff = TEMP - CRITT
    IF 'yes' .IN. NEGSZERO
        CALCULATE Tempdiff = Tempdiff * (Tempdiff .GT. 0)
    ENDIF
    CALCULATE Lastday = NVALUES(TEMP)
    VARIATE [VALUES = 1...Lastday] Day
    CALCULATE MEANDIFF = CUM(Tempdiff) / Day
    CALCULATE Exceed = Tempdiff .GT. 0
    IF 'yes' .IN. NUMBER
    CALCULATE NPERIODS = SUM(SHIFT(Exceed) .GT. Exceed) + \
        Exceed$[Lastday]
    ENDIF
    IF 'yes' .IN. DURATION
        CALCULATE DPERIODS = SUM(Exceed)
    ENDIF
    IF 'Tempdiff' .IN. PRINT
        PRINT [IPRINT = *] \
            'Cumulative mean temperature differences from', \
            CRITT,'degrees'
        & ' Day Mean'
        & [IPRINT = *] Day,Tempdiff; FIELDWIDTH = 4,5
    ENDIF
    IF 'number' .IN. PRINT
        PRINT [IPRINT = *] \
            'Number of periods above critical temperature:',NPERIODS
    ENDIF
    IF 'duration' .IN. PRINT
        PRINT [IPRINT = *] \
            'Duration of periods above critical temperature:',DPERIODS
    ENDIF
ENDPROCEDURE
SCALAR Grwthmin; VALUE = 5
VARIATE [VALUES = 0,2,3,6,7,5,2,8,9,4] Sprngtmp
HEATUNIT [NEGSZERO = yes; NUMBER = yes; DURATION = yes; \
    PRINT = tempdiff,number,duration] TEMP = Sprngtmp; \
    CRITT = Grwthmin
HEATUNIT [PRINT = nothing] TEMP = Sprngtmp; CRITT = Grwthmin; \
    MEANDIFF = Meandiff
PRINT Meandiff
STOP
```

Chapter 10

Exercise 10(1)

```
CALCULATE Random = URAND(83430; 1000)
CORRELATE [MAXLAG = 50; GRAPH = auto] Random
TSM Ar[1]; ORDERS = !(1,0,0)
ESTIMATE Random; TSM = Ar[1]
STOP
```

Exercise 10(2)

```
READ [SETNVALUES = yes] Temp
    (data)
CALCULATE Maxday = 31 + NVALUES(Temp)
VARIATE [VALUES = 32...Maxday] Day
CALCULATE Dtemp,DDtemp = DIFFERENCE(Temp,Dtemp)
GRAPH Temp; Day
& Dtemp; Day
& DDtemp; Day
    " Choose first differences "
TSM Ima[1...3]; ORDERS = !(0,1,1),!(0,1,2),!(0,1,3); \
    PARAMETERS = P[1...3]
ESTIMATE Temp; TSM = Ima[1]
TKEEP RESIDUALS = Res[1]
    " Use parameters of first model as starting value for second "
VARIATE [VALUES = #P[1],0] P[2]
ESTIMATE Temp; TSM = Ima[2]
TKEEP RESIDUALS = Res[2]
VARIATE [VALUES = #P[2],0] P[3]
ESTIMATE Temp; TSM = Ima[3]
TKEEP RESIDUALS = Res[3]
    " Choose IMA(2) "
CORRELATE [MAXLAG = 25; GRAPH = auto] Res[2]
STOP
```

Exercise 10(3)

```
READ [SETNVALUES = yes] Temp
    (data)
READ [SETNVALUES = yes] Tempadd
    (data)
CALCULATE N[1,2] = NVALUES(Temp,Tempadd)
VARIATE [VALUES = #Temp,#Tempadd] Temp
```

```
TSM Ima[2]; ORDERS = !(0,1,2)
ESTIMATE Temp; TSM = Ima[2]
VARIATE [VALUES = #N[1](0),#N[2](1)] Step
TRANSFER Step
ESTIMATE Temp; TSM = Ima[2]
STOP
```

Chapter 11

Exercise 11(1)

```
CALCULATE Noise = NED(URAND(77529; 100))
& Trend = Noise + !(0.1,0.2 ... 10)
CORRELATE [MAXLAG = 10] Noise,Trend; AUTO = Nauto,Tauto
FOURIER Nauto,Tauto; TRANSFORM = Nspect,Tspect
PRINT Nspect,Tspect
STOP
```

Exercise 11(2)

```
READ [SETNVALUES = yes] Average
     98.7    99.1    93.2    97.8   100.2    97.5   101.3
    101.3    99.6   107.8   101.4   101.4   100.2   100.9
    101.1   102.5   101.0    98.7    99.9    96.7 :
CORRELATE [GRAPH = auto; MAXLAG = 10] Average; AUTO = Acf[1]
FOURIER [PRINT = transform] Acf[1]; TRANSFORM = Spect[1]
CALCULATE X = 2*3.14159*!(1...20)/20
& B = 2/10*SUM(Average*SIN(X))
& C = 2/20*SUM(Average*COS(X))
& Adjave = Average - (B*SIN(X) + C*COS(X))
CORRELATE [MAXLAG = 10] Adjave; AUTO = Acf[2]
FOURIER Acf[2]; TRANSFORM = Spect[2]
PRINT Spect[]
STOP
```

Exercise 11(3)

```
VARIATE [VALUES = 1...55] Month
UNITS Month
READ Call
    (data)
CORRELATE [MAXLAG = 24] Call; AUTO = Acf[1]
```

```
CALCULATE I = !(0...24)/25
& Parzen = 2*(1 - I)**3*(I >= 0.5) + (1 - 6*I*I*(1 - I))*(I < 0.5)
& Tukey = (1 + COS (3.14159 + I))/2
& Acf[2,3] = Acf[1] * Parzen,Tukey
FOURIER Acf[]; TRANSFORM = Spect[1...3]
PRINT Spect[]
" The frequencies are: 0, 1/48, 2/48, 3/48, 4/48 = 1/12, ..."
CALCULATE Nat12,Parz12,Tukey12 = Spect[]$[5]
PRINT Nat12,Parz12,Tukey12
STOP
```

References

Anderson, O.D. (1976). Time series analysis and forecasting. Butterworths, London.

Berman, M. and Culpin, D. (1986). The statistical behaviour of some least squares estimators of the centre and radius of a circle. Journal of the Royal Statistical Society B 48, 183–96.

Bliss, C.I. (1970). Statistics in biology, Volume II. McGraw–Hill, New York.

Boer, J. and Jansen, B.C.P. (1942). Studies on the nutritional value of butter in comparison with other fats supplemented with various amounts of Vitamins A and D. Archives Néerlandais de Physiologie de l'Homme et des Animaux 26, 1–177.

Box, G.E.P., Hunter, W.G., and Hunter, J.S. (1978). Statistics for experimenters. Wiley, New York.

Box, G.E.P. and Jenkins, G.M. (1970). Time series analysis, forecasting and control. Holden–Day, San Francisco.

Buchan, A. (1890). Meteorology of Ben Nevis. Transactions of the Royal Society of Edinburgh, Volume 34.

Bulmer, M.G. and Perrins, C.M. (1973). Mortality in the Great Tit *Parus major*. Ibis 115, 277–81.

Chatfield, C. (1980). The analysis of time series: an introduction (2nd Edition). Chapman and Hall, London.

Davies, O.L. (1967). The design and analysis of industrial experiments. Oliver and Boyd, London.

Delaney, M.J. and Healy, M.J.R. (1966). Variation in white-toothed shrews in the British Isles. Proceedings of the Royal Society B 164, 63–74.

Digby, P.G.N. (1986). Graphical displays for classification. In: To pattern the past! (ed. A. Voorips and S.H. Loving). PACT, Strasbourg.

Doran, J.E. and Hodson, F.R. (1975). Mathematics and computers in archaeology. Edinburgh University Press.

Gabriel, K.R. (1971). The biplot graphic display of matrices with applications to principal components analysis. Biometrika 58, 453–67.

Gabriel, K.R. (1981). Biplot display of multivariate matrices for inspection of data and diagnosis. In: Interpreting multivariate data (ed. V. Barnett). Wiley, Chichester.

Genstat 5 Committee (1987). Genstat 5 reference manual. Clarendon Press, Oxford.

Gomes, F.P. (1953). The use of Mitscherlich's regression law in the analysis of experiments with fertilizers. Biometrics 9, 498–516.

Gower, J.C. (1971). A general coefficient of similarity and some of its properties. Biometrics 27, 857–72.

Hartigan, J.A. (1975). Clustering algorithms. Wiley, New York.

Heyes, J.K. and Brown, R. (1956). Growth and cellular differentiation. In: The growth of leaves (ed. F.L. Milthorpe). Butterworth, Glasgow.

Hollingshead, A.B. (1946). Elmtown's Youth: the impact of social classes on adolescents. Wiley, New York.

Jenkins, G.M. and Wilkinson, G.F. (1982). The estimation of a change in price structure in the U.S. telephone industry. In: Case studies in time series analysis, Volume 1 (ed. G.M. Jenkins and G. McLeod). Gwilym Jenkins and Partners, Lancaster.

Johnston, A.E., Lane, P.W., Mattingly, G.E.G., Poulton, P.R., and Hewitt, M.V. (1986). Effects of soil and fertilizer P on yields of potatoes, sugar beet, barley and winter wheat on a sandy clay loam soil at Saxmundham, Suffolk. Journal of Agricultural Science 106, 155–67.

Keyfitz, N. and Flieger, W. (1972). Population: facts and methods of demography. Freeman, New York.

Kipling, C. and Roscoe, M.E. (1977). Surface water temperature of Windermere. Freshwater Biological Association, Ambleside.

Lane, P.W., Galwey, N.W., and Alvey, N.G. (1987). Genstat 5: an introduction. Clarendon Press, Oxford.

Lindley, D.V. and Scott, W.F. (1984). New Cambridge elementary statistical tables. Cambridge University Press.

Loxdale, H.D., Tarr, I.J., Weber, C.P., Brookes, C.P., Digby, P.G.N., and Castanera, P. (1985). Electropheretic study of enzymes from cereal aphid populations. III. Spatial and temporal genetic variation of populations of *Sitobion avenae* (F.) (Hemiptera: Aphididae). Bulletin of Entomological Research 75, 121–41.

Luo, S., Liang, S., Ye, S., Yan, S., and Li, Y. (1977). Analysis of periodicity in the irregular rotation of the earth. Chinese Astronomy 1, 221–7.

McCullagh, P. and Nelder, J.A. (1983). Generalized linear models. Chapman and Hall, London.

Mardia, K.V., Kent, J.T., and Bibby, J.M. (1979). Multivariate analysis. Academic Press, London.

Mourant, A.E., Kopeć, A.C., and Domaniewska-Sobczak (1976). The distribution of the human blood groups and other polymorphisms (2nd Edition). Oxford University Press.

Ratkowsky, D.A. (1983). Nonlinear regression modeling. Dekker, New York.

Scheffé, H. (1959). The analysis of variance. Wiley, New York.

Sherlock, P.L., Bowden, J., and Digby, P.G.N. (1985). Studies of elemental composition as a biological marker in insects. IV. The influence of soil type and host-plant on elemental composition of *Agrotis segetum* (Denis & Schiffermüller) (Lepidoptera: Noctuidae). Bulletin of Entomological Research 75, 675–87.

Siegel, S. (1956). Nonparametric statistics for the behavioral sciences. McGraw–Hill, Tokyo.

Snedecor, G.W. and Cochran, W.G. (1980). Statistical methods (7th Edition). Iowa State University Press.

Thom, A. (1979). Megalithic sites in Britain. Oxford University Press.

Tunnicliffe Wilson, G. (1982). Time series in Genstat. University of Lancaster.

Villars, D.S. (1947). A significance test and estimation in the case of exponential regression. Biometrics 18, 596–600.

USDA (1966). Consumer expenditure survey report, Numbers 31–34. Agricultural Research Service, United States Department of Agriculture.

Index

This index contains entries for all items in the Genstat language discussed in this book, together with references to English words and phrases. Entries have been combined where a Genstat item is also an English word; so, for example, references for calculation are found under CALCULATE. Option and parameter names are given only under the entries for each directive. No references have been made to the content of examples or exercises except where new information about Genstat has been presented.

QMW LIBRARY
(MILE END)